Noguchi and His Patrons

Hideyo Noguchi, 1920

Noguchi
and His Patrons

Isabel R. Plesset

Rutherford • Madison • Teaneck
Fairleigh Dickinson University Press
London and Toronto: Associated University Presses

© 1980 by Associated University Presses, Inc.

Associated University Presses, Inc.
Cranbury, New Jersey 08512

Associated University Presses
Magdalen House
136-148 Tooley Street
London SE1 2TT, England

Associated University Presses
Toronto M5E 1A7, Canada

Library of Congress Cataloging in Publication Data

Plesset, Isabel Rosanoff, 1912-
    Noguchi and his patrons.

    Bibliography: p.
    Includes index.
    1. Noguchi, Hideyo, 1876-1928. 2. Microbiologists—
Japan—Biography. I.   Title.
QR31.N6P55   616.01'4'0924 [B]   78-66819
ISBN 0-8386—2347-6

PRINTED IN THE UNITED STATES OF AMERICA

2255140

Contents

5

List of Illustrations

Preface and Acknowledgments

WHO WAS HIDEYO NOGUCHI, WHY SHOULD HIS BIOGRAPHY BE WRITTEN AT this time, and how did I develop such an interest in this man? These questions have been asked of me repeatedly during the eight years of my research and preparation of this book.

When Noguchi was born in Japan in 1876, his family were impoverished peasants living in an isolated mountain village. From the time of the mutilation of his left hand in a fire in his infancy until his dramatic death of yellow fever in West Africa at the age of fifty-one, Noguchi's lonely life of hardworking self-tutelage was spiced with audacious episodes, all in furtherance of his grandiose ambition to rid himself and his family of the degradation of their lives. By the time he was thirty-five years old, Noguchi was one of the world's most eminent microbiologists, a member of The Rockefeller Institute for Medical Research in New York, destined — it seemed — to free the world of one infectious disease after another as he isolated their causative organisms and made vaccines to combat each organism. Less than twenty years later he was dead, a victim not only of one of the infectious diseases on which he had been working but a victim as well of his own error in his work with that disease, yellow fever.

From his position as one of the most distinguished microbiologists of his day, Noguchi rapidly became one of the most controversial. His work on yellow fever precipitated a controversy over his whole career and even the circumstances of his death. Did he commit suicide? There was even a suggestion that he had been murdered. Although biographies, novels based on Noguchi's life, sketches, and commentaries have appeared, not frequently but steadily, ever since his death, Noguchi's lasting work is relegated to the obscurity of inner pages of textbooks in the West, and is rarely mentioned in scientific literature in Japan. His name is not well-known today in the West, but he has become a folk hero in Japan. Every person who has attended elementary and junior high school in Japan knows the story of Noguchi, the martyr to science, much as American children learn the story of George Washington and the cherry tree.

Material on Noguchi's life and work has been destroyed, withheld, and distorted on both sides of the Pacific Ocean. Although some of it is no longer recoverable, much remains, and the emotional barriers to its disclosure have begun to soften. Clearly it is time for the truth to be written.

My father was the clinical director of one of the New York State psychiatric hospitals at the time when Noguchi's initial work with the organism of syphilis was in progress. He supplied Noguchi with pathologic material for study, and Noguchi helped set up a clinical and research laboratory in that state hospital. My father and Noguchi were friends at this time, which was before I was born, and I grew up hearing stories about Noguchi. On my first visit to Japan in 1968 I naturally sought out the Noguchi memorials, where I developed a new interest in learning more about Noguchi's life in the West. What started as a vague promise on my part to search for such information grew into this study.

Previous writers on the subject of Noguchi have all necessarily been limited in their scope, but their works have all been useful to me. I have drawn especially on the work of Tsurukichi Okumura, *Noguchi Hideyo* (1933), which includes a carefully edited collection of Noguchi's letters to his friends and family in Japan as well as a narrative account of his family, childhood, and life in the Orient before coming to the United States in December 1900. Any biographical material that is not specifically attributed to another source was drawn from Dr. Okumura's work. Gustav Eckstein's biography of Noguchi (1931) has been the most generally accepted work in the English language, but Eckstein was almost as limited in his access to material in the West as Okumura was.

Important new sources of information are now available from the papers of Noguchi's Western colleagues, which have appeared in various archives in recent years. Most notably, the papers of Simon Flexner, Peyton Rous, Harold Amoss, and Peter Olitsky are all in the Library of the American Philosophical Society, where I was permitted to study them. Equally valuable sources of material are now available in the archives of the Rockefeller Foundation, the Rockefeller Institute, and the Rockefeller family archives, all of whom granted me access to the material collected there. The Noguchi Memorial Foundation in Tokyo has been collecting and verifying information ever since its establishment in 1937, and I was supplied with whatever material I requested from this source.

In the study of this newly available material combined with the older sources of information, my objective has been to understand and reconcile the widely divergent views of Noguchi, and to examine these views in the perspective of the changing times in which he lived and worked. In the service of this objective, I have found and communicated with more than fifty people who knew Noguchi either as a colleague or as a personal friend. I have had an equal number of contacts with members of families of Noguchi's associates, and with a number of scientists who have some special knowledge of the surroundings and problems with which Noguchi lived and worked. Dr. Fred Soper worked on yellow fever in Brazil after Noguchi had been there; Dr. J. Austin Kerr followed Noguchi not only in Brazil but in West Africa as well; Dr. David Weinman not only made a careful analysis of Noguchi's work in Oroya fever and verruga peruana but also repeated that work in Peru. Dr. Max Theiler was awarded the Nobel Prize in 1951 for his development of an attenuated yellow fever virus that was suitable for use as a preventive vaccine.

Special insights have come to me from the help of special people. Hisarharu Tsukuba, a Japanese writer on science who wrote a modern view of Noguchi, discussed the subject with me. Dr. L. Takeo Doi, psychiatrist at Tokyo University School of Medicine, helped me with cultural interpretations of some of Noguchi's letters. Professor Theodore Y. Wu spent many hours with me in conveying the mood, the allegory, and the cultural and literary background of Oriental forms of expression. I am equally indebted to Dr. Saul Benison, Dr. Martin Bronfenbrenner, Dr. Max Delbrück, Dr. Rebecca Lancefield, Dr. Philip McMaster, Dr. Philip Reichert, Dr. Cornelius B. Philip, and Dr. Evelyn Tilden.

For the general understanding, analysis, and reporting of technical material, I have had the valuable assistance of Dr. Marjorie L. Plesset, a physician; Dr. Judith E. Plesset, a biochemist; and Dr. Howard Winet, a biophysicist who had no family obligation to come to my aid. Dr. Fawn Brodie, a biographer and historian, has read my manuscript and made valuable and constructive suggestions. Translation of Noguchi's letters and other Japanese material was done for me by Mrs. Masako Ohnuki, with the assistance in special instances of Professor Rokuro Muki and Mr. Toichi Hasatani. I have not included a bibliography of Noguchi's published works because it would only duplicate the presentation of Dr. C. E. Dolman in his article on Noguchi in *Dictionary of Scientific Biography*. Since that dictionary is so readily available, I refer the reader to it.

Dr. Frederick Seitz, formerly president of The Rockefeller University, arranged for my access to the various archives where Mrs. Sonia Mirsky, Mrs. Ruth Sternfeld, Mr. Warren Hovious, and Dr. Joseph Ernst guided me though the enormous volume of material. Mr. Murphy Smith gave me similar assistance in the Library of the American Philosophical Society, and Mr. Hideo Sekiyama provided me the same competent courtesy at the Noguchi Memorial Foundation in Tokyo. I must also thank Father Joseph

T. Durkin, S.J., for granting me access to the Carrel Papers at Georgetown University, the New York Academy of Medicine for access to the Noguchi letters to Famulener, and the University of Pennsylvania for the picture of Noguchi taken in Copenhagen in 1904. I give my special thanks to Mr. Kiyoshi Saito for permission to use his Snow Country print in the design of the book jacket.

I express my appreciation to Mr. James Thomas Flexner for permission to quote from the unpublished writings of his father, Dr. Simon Flexner. I am also grateful to Dr. David Kligler, Dr. Sten Madsen, and Mr. S. Kobayashi for providing me with material from their family papers.

Far beyond the call of duty, Mr. Roderick Casper and Miss Sophia Yen of the Millikan Library of the California Institute of Technology responded to the scores of requests I made of them. I am grateful to Mrs. Frances Larson for editorial help.

I wish to thank the following for having given me permission to quote from copyrighted material:

Appleton-Century-Croft for permission to quote from *Microbiology,* edited by David T. Smith, Norman F. Conant, John R. Overman.

Little, Brown and Company for permission to quote from *As I Remember Him: The Biography of R.S.* by Hans Zinsser.

The John Hopkins University Press for permission to quote from "Hideyo Noguchi 1876-1928" by Paul Franklin Clark.

The University of Chicago Press for permission to quote from *The Newer Knowledge of Bacteriology and Immunology* edited by Jordan and Falk.

American Journal of Tropical Medicine and Hygiene for permission to quote from "A History of Bartonellosis (Carrión's Disease)" by Myron G. Schultz.

World Health Organization, Geneva, Switzerland and Dr. R. R. Willcox, London, for permission to quote from *Treponema Pallidum* by R. R. Willcox and T. Guthe.

Yale University Press for permission to quote from *A History of Poliomyelitis* by John R. Paul.

Harcourt Brace Jovanovich for permission to quote from *The Sweeping Wind* by Paul de Kruif.

The Rockefeller Institute Press for permission to quote from *A History of The Rockefeller Institute 1901-1953: Origins and Growth* by George W. Corner.

It has been my very special and great good fortune that Professor Lawrence Olson of Wesleyan University, Middletown, Connecticut, has taken an interest in this work. With his comprehensive knowledge of Japan, its language, its culture, and its people, he has guided and encouraged this project. Only thus could it have been done.

And finally to my husband, Milton, who has now seen many places he never intended to see—Inawashiro, Woodlawn Cemetery in The Bronx, Aizu-Wakamatsu, and the waiting rooms of countless libraries, offices, and archives—thank you.

Introduction

IN THE COLD WET SPRING OF 1878, SHIKA NOGUCHI WAS THE ONLY MEMBER of her family left to prepare the paddy fields for planting the rice, but she knew she could do it. She had already done it more than once in the grim twenty-four years of her life, and this year she had a greater incentive than ever before. She had given birth to a son — the first boy born in her family since her grandfather — and she had brought the baby safely through the first two perilous winters of its infancy.

Even in her childhood, Shika had found whatever strength she needed in dreams of restoring her family to the position it had once held in the village in the days of her great-grandfather. With little but her own efforts she had held on to the family farm. She had married a man who had accepted adoption into the Noguchi family, and now that her son was seventeen months old, strong and healthy, she could almost see him as the head of a prosperous household and the family treated with respect in the village.

Like many other peasant women working in the fields, she made a practice of swaddling her baby in a straw hamper, or *ejiko*, and leaving him in the warm house beside the open fire pit commonly found in Japanese homes. Like all of the mothers she knew, Shika expected that the many strings that bound her baby son in his covers were secure enough so that he could not work himself free even when he awoke, restless and hungry. Moreover, the baby's grandmother was there, aging and deaf though she was, and his four-year-old sister could also keep watch over him.

Shika customarily returned to the house after work, stirred up the fire, and hung a pot of water over it for soup, and then went out to a storage pit behind the house to fetch vegetables. One day she heard sudden screams from the baby and dashed back into the house, where she found that he had gotten loose and crawled into the fire. He was burned over much of his body and his left hand was horribly damaged. Shika sought no medical help, but drew on her own resources and the advice of neighbors for treatment. She made dressings which kept the right hand open and she worked with it, leaving the left hand folded down upon itself, apparently seeing at once that she could do nothing to save it. An infection developed and the baby was very ill, but somehow he survived. After the burns had healed, the baby's left hand was entirely useless. The end joints of the fingers were gone, the remaining joints had adhered to each other to form a solid clump, and the thumb was drawn down to the wrist and had become attached to it.

According to village legend, Shika went without sleep for three weeks in order to care for her baby, propping her eyelids open with little pieces of wood to keep awake. After the child recovered, she took him with her wherever she went; she carried him on her back as she moved from place to place and suspended him from a tree in his straw hamper while she worked in the fields. She vowed to her neighbors that she would devote her life to the future of this child whose mutilation she felt she had needlessly permitted by leaving him in the care of the old lady and the little girl. The feeling of guilt on the part of the mother of a seriously injured child is persistent and profound, no matter how little that feeling may be deserved. If Shika had been more careless than we have been told, the facts have been lost in the telling. The villagers may have tried to relieve her of some of her feeling of responsibility as they told their stories of the accident, recognizing that such an accident could have happened to any one of them since they cared for their babies in the same way.

During the weeks of her little son's illness and in the years she carried him with her, Shika questioned whether he could ever become a successful farmer with such a deformity. She fished at night, cleaned the fish to sell in the morning, and continued to work in her fields. But with every young rice plant she would put into the ground, Shika pondered the implications of her son's handicap. The dreams, the fantasies of what her family had once been, seemed useless now. But she knew nothing beyond life in her tiny village and she could search for answers only in the stories of her family background and in her own brief but intense experience with life. At this time of her greatest tragedy, she had nowhere to turn but to the past.

Noguchi and His Patrons

1

Origins

In theory the farmer ranked next to the samurai and above the
artisan and the trader. But in practice the men who tilled the soil
were heavily oppressed, and their life was often wretched. Honda
Masanobu, Ieyasu's trusted advisor, wrote that the peasant was the
foundation of the state and must be governed with care. He must be
allowed neither too much nor too little, but just enough to live on
and to keep for seed in the following year. The remainder must be
taken from him as tax.
— George Sansom, *A History of Japan 1615—1857*[1]

THE HISTORY OF THE NOGUCHI FAMILY, INCLUDING THE CHILDHOOD OF
Hideyo Noguchi, is known only from such a mixture of peasant tales with
village records that it takes on the air of a legend from a far-off time. The
family first appears in tax records for the year 1797 when their village,
Sanjōgata, was hardly more than a cluster of tenant farmers, about thirty
households in all. The Nihei and the Noguchi families are said to have
been the most prominent and well off, but there had been so much inter-
marriage between them that it has not been possible to trace the families
back to their separate origins or to learn how either family gained permission to use a family surname.

Sanjōgata was one of the villages in the domain of the Aizu clan, a fief
that had been granted in the earliest days of Japan's consolidation under a
military dictatorship in the first years of the seventeenth century. The Aizu
territory included mountain areas as well as farmlands in the *Tōhoku* (north-
east) district of Honshū, Japan's main island. Because of its rigorous

19

climate and mountainous character, the *Tōhoku* district was more difficult to develop than the southern part of the country and thus progressed
more slowly; in the eleventh century it was referred to in literature as
michinoku, the back country, a label that it was borne into modern times.

Japan's emperor has almost always been a symbolic figure. Power is
wielded on his behalf by others to whom he ceremonially delegates it. At
the beginning of the seventeenth century, the Tokugawa family (clan)
achieved a dominant position among the feudal lords who had been war-
ring among themselves for generations. Granted imperial recognition, the
Tokugawa established the dictatorship that governed Japan from 1603 to
1867, and the shogun, the dictator, was chosen by this family from within
its own ranks. Local lords held their fiefs at the pleasure of the shogun;
although substantial demands were put upon them through their obligations to the Tokugawa, within their own domains they ruled as they saw fit.
The original holder of the Aizu fief was a kinsman of Ieyasu, the first
Tokugawa shogun. Kinsmen passed on to their descendants the prestige of
this original relationship along with the family surname, Matsudaira,
granted to all kinsmen lords.

Thus the Aizu lord had the power of prestige to back up his rule, but the
nature of his domain required toughness and capability as well, a
challenge that the succeeding generations of Aizu lords were well able to
meet. They set high value on cultural development and education, which
they provided for their samurai, and the Aizu bureaucracy functioned so
smoothly and efficiently that it has been described as having the beauty of
a work of art.[2]

This beauty may well have escaped the notice of the Aizu farmers.
Following the so-called Seven Year Famine of Tenmei, a disaster during
which it has been estimated that throughout Japan "at least a million people must have died of famine or pestilence within less than a decade,"[3] the
twofold response of the Aizu lord showed little compassion. He merely
ordered a reassignment of land among his farmers and raised the level of
yield that the villagers were ordered to produce in line with his policy of
keeping production as steady as possible. Such were the conditions under
which the Noguchi were honored with a land assignment; they didn't
flourish for long.

According to Noguchi's Japanese biographer, Tsurukichi Okumura, the
Noguchi family were considered good farmers. Yet during the first fifty
years of the tax record they declined steadily from full payment of a slightly
higher-than-average assessment to debtors to the community reserves. The
Noguchi family had to borrow rice in 1846, and continued to borrow in the
following years. At this time the family consisted of Iwakichi, the head of
the house, his wife Mitsu, and two daughters, aged sixteen and nine. In
1848, the younger girl was released for adoption. It is possible that in this

instance "adopted" was a euphemism for "bought," and that the funds the family received in payment for their younger daughter enabled Iwakichi to adopt a son for marriage to the older daughter. The couple were married in 1851, but when a girl was born to them in 1854, the family members felt that they could no longer survive as a farming unit and began to disperse. The baby girl, who was to become the mother of Hideyo Noguchi, was considered at the time of her birth to be the family's final catastrophe.

Iwakichi secured an appointment in the castle of the lord at Wakamatsu, about five miles from Sanjōgata. The adopted son, father of the little girl, simply disappeared and for a while the family did not know where he was. Even the baby's mother left to take employment in a household of another village, leaving only Mitsu to care for her little grandchild.

For the next four or five years, these two lived together in extreme poverty. The child, who had been given the name Shika (the deer), learned in these years barely beyond her infancy to help her grandmother produce a small amount of rice on the farm. The indebtedness of the family had become so great that it could borrow only a tiny percentage of its needs from the community. The woman that Shika was to become was the child of those years of deprivation. One wonders how Mitsu could have nourished the spirit and stamina of her little granddaughter to face this kind of life. Like many of the farm women of the area even today, Mitsu had sustained herself in part through faith in Kannon, the goddess of Mercy, and we know that she passed this faith on to Shika. Was it also Mitsu who filled the child's head with tales of the previous glory of the family and dreams of restoring the family to a better life?

Old people of the village told Okumura that when Shika was five years old, she heard a rumor that her father was serving in the house of a high level Aizu vassal who lived in Inawashiro, a larger village than Sanjōgata about four miles from the Noguchi farm. Shika walked to this house by herself, and when she approached the gate she walked back and forth in front of it, which was "a fearsome thing for even an adult to do." A guard questioned her and recognized that she was the daughter of a footman who went by the name of Seisuke in this household. The guard took Shika inside to wait for her father who was out doing errands, but the master's wife heard that she was there and spoke with her. The lady was touched by the little girl's courage, gave her some presents, and when Seisuke returned the lady granted him leave to go home for a visit. Seisuke was able to stay at his home for a few weeks, but had to return to his job.

When Mitsu became too ill to work, Shika was taken on as an "employee" in one of the Nihei households. Some three years later, when Shika was nine years old, Mitsu died and the child's grief was profound. The villagers remembered her clinging to the old lady's body and crying inconsolably. The priest of the temple was astonished shortly afterward when Shika came with a contribution of ten sen* for the temple. He accepted her tiny offering as an adult gesture and symbol of her need for consolation.

When Shika was eleven years old, her mother returned after an absence of nine years, and the family decided to reunite and try to make a new start. Iwakichi and Seisuke had just finished their periods of service, and Shika was released by the Nihei. But there was "an insoluble bitterness" in the household, trouble started almost at once, and Seisuke left again.[4] For Shika, this meant only that once again she had lost her father, and that once again she was left with adults who were worn out. Iwakichi did not have the strength left for farming and Shika's mother had never been very effective, so the household burdens fell on the only robust member of the family, the spirited Shika.

During the first winter they were together, there was a heavy snowfall. Before the snow could be removed from the roof there was a second storm, with the result that the roof caved in and the whole house collapsed. They lived under mats strung over the patched-up frame of the house for the rest of the year and through the entire next winter. No one was surprised when Iwakichi died during this period at the age of sixty-one; the last funds that the family could scrape together went for his funeral. Now even Shika recognized the futility of any efforts to hold the family together and she went to work in another of the Nihei households.

This family, who lived across the road from the Noguchi house, seem to have recognized that Shika needed the protection and support of a stable family, particularly in the light of much larger events which had been taking place throughout Japan and which were reaching their climax at this time.

For more than two hundred years the shogunate had isolated Japan from all but the most highly restricted contact with the world beyond its islands, but this isolation was brought to an end when American warships entered Tokyo Bay in 1845.[5] Furthermore, the system within the country was weakening under the strains of its lopsided allocation of resources; although the burden of the rice economy had been carried by the labor of the farmers, its benefits had been dissipated in the support of the samurai. The objective of the system had been peace within the country and this had been achieved; but the samurai had proliferated. Many of the local lords had become bankrupt, not only from supporting their own samurai but also from maintaining the lavish establishments required by their obligations to the shogunate. These conditions had created a whole class of wandering samurai with no master to serve (*rōnin*), and had allowed the merchant class, the lowest on the social scale, to quietly expand its own subculture

---

»100 sen = 1 yen, about 50 cents.

Noguchi Family Home, Inawashiro, Japan. Hideyo Noguchi Memorial Association

and to hold much of the economic power of the country in its hands.

That changes must be made was clear, but the direction that these changes should take was hotly contested. Contention was brought to focus on the issues of whether to open the country to trade and to Western style industrialization, but the real issue was which of the many contending factions would hold control of the imperial court with enough strength to put its own policies into force.

The shogunate itself was threatened and the Aizu lord was naturally expected to commit all his resources — political, economic, military, and spiritual — to the defense of his lord. He did so with a tenacity that has guaranteed the Aizu people a permanent place in the saga of Japanese heroes. The Tokugawa forces met their final defeat in Tokyo in the fall of 1867; power was restored to the emperor, but the Aizu fought on in their own domain against the armies now designated as the Imperial forces. The Aizu surrender came only with the burning of the castle of Wakamatsu in November 1868, an event whose poignancy was heightened for the Aizu by the suicides of nineteen young samurai, all under the age of seventeen, who had vowed to defend the castle with their lives.

In such times, when loyalty was meeting its true test, all the time-honored virtues were reaffirmed; loyalty to one's lord and to family, sincerity in meeting obligations, and working to the limit of one's capacity. Those times gave to Shika aspirations that had never been known to her grandmothers. She could not only throw herself into becoming a living symbol of the virtuous woman, devoted to the welfare of her family, she could also expect an acknowledgment of her efforts that previous generations might well have taken for granted. Thus, the facts of her childhood heroics may have been somewhat embellished, particularly after her son achieved his worldwide distinction; as the villagers told the stories of their heroine, they understated the circumstances of the times and the help she was given by her neighbors. Although it is not easy to separate fact from myth in the stories of Shika's early childhood, the facts appear more clearly after she started to work for the Nihei family in 1867.

Renzaburo Nihei was the headman of Sanjōgata, a position for which an election was held, but one that was passed on so regularly from father to son that it was almost hereditary. The headman represented the farmers of his village in their dealings with the officers of the Aizu lord; he kept the tax records and the production records, and it fell to him to present the farmers' grievances, which was often a risky thing to do. The headman was respected and popular with the villagers, and thus Nihei was in a position to be of significant help to Shika. At one time she was given a cash award by the Aizu government office; her efforts had been brought to official attention and she was cited for her faithfulness in carrying out her family obligations.[6]

As the fighting between the Aizu and the Imperial forces drew closer, soldiers were stationed in the village and were housed in the temple. This may have been the time at which the Nihei family decided to take Shika under their protection, for Shika had shown that she was interested in more than just her family and the farm. During the time her family was living in the roofless hovel, she had tried to learn to dance as she saw other girls do even though she had to appear in *tabi* (Japanese socks) made of paper instead of cloth.

During the five years she spent in the Nihei household, Shika made substantial improvement in the level of her material resources. She took advantage of a period when the village was cut off by the fighting, and borrowed some money that she used to buy material and made matches that she then sold in the village. During her employment in the Nihei house, she concentrated on the collection of material for the repair of her family house. After the fighting was over, Nihei hired a carpenter to do some work on his house. Shika was ready with her materials and she persuaded Nihei to have the carpenter repair her family house as well. In 1872, when she was eighteen years old, a friend of Nihei's acted as go-between to arrange a marriage for Shika as well as for the adoption of her bridegroom into the Noguchi family. Shika had indeed accomplished, albeit with some help, one essential step toward the perpetuation of the family that had regarded her birth as their ultimate disaster.

It was common practice for a family to arrange for the adoption of a second or later born son of an acceptable family for marriage to the daughter of a family who either had no son at all or no son they considered capable of carrying on the family responsibilities. It is through this custom that the Japanese can boast today of their imperial line being the oldest in history. In higher circles there was always a certain *quid pro quo* in this arrangement; in addition to the genetic outcome there were cultural, socioeconomic, and political considerations. For those further down the social ladder, these considerations were harder to define, and the role of an adopted son had less to offer a young man and was avoided whenever possible.

Shika had little to offer; she was neither rich nor pretty. The very impudence of her expecting a man to give up his own family for the dubious honor of becoming head of the Noguchi family in its current condition did little to recommend her as a compliant Japanese wife. She must have put forth considerable effort to persuade Nihei to cooperate in finding a man who would accept her conditions. It is not surprising, therefore, that so little has been told of the man who married Shika. It seems reasonable to doubt that she ever truly acknowledged her husband as head of her family or that she ever intended that he should assume that position. Her grandfather had abandoned his efforts to keep the family farm going. Her father, also an adopted son, had never assumed the role of head of the household; the efforts she had made as a little girl to get her father to remain at home and take charge met with failure. She had assumed the role of head of the household at a tender age. It is likely that her attitude toward men was determined by what she saw as the weakness of her father and grandfather, and that this attitude colored her relationship with her husband before she ever met him.

Before he was adopted, her husband's name was Sayosuke Kobiyama. He was the oldest son of a family from another village, and was thirty years old at the time of the marriage. Hideyo Noguchi's American biographer, Gustav Eckstein, speaks of Sayosuke as a man who loved *sake* just as his father before him had loved the wine so much that he had mortgaged the family property "all for drink" and then gone off to Hokkaidō.[7] Japanese biographers write that he did nothing but drink. The reputation of the Kobiyama family men as drinkers probably explains why the oldest son of this family was released for adoption into any other household, much less one with as little promise as the Noguchi family. No one wanted Sayosuke as a husband.

Immediately following the fall of the shogunate, there was a period during which police protection had not yet developed and throughout Japan a special kind of gang assumed a sort of police function in local areas. Supported by gambling, influence peddling, probably prostitution, and various kinds of criminal activity for hire, these gangs functioned much like the original Mafia. A family-style group with close inner loyalties and intense intergang rivalries, these gangs were looked to by ordinary helpless citizens for assistance and protection. Men known as *gunpu*, who had performed menial chores for the army during the period of fighting, were admitted to these gangs as lower level members. Sayosuke was one of such *gunpu*.

The boss, or gang leader, called *oyabun*, was looked up to with awe and respect. Although gambling was frowned upon by upper-class people as wasteful, it was not illegal; it was considered a legitimate means of support for the gangs in return for the service they rendered. In true gang tradition, territorial rights were agreed upon, or decided by fighting between the gangs, and each *oyabun* appointed local bosses to the sections of his territory. Recent popular movies and stories in Japan still portray episodes in the lives of these gangsters who retain their glamor for ordinary folk.

Although Sayosuke's occupation was respectable, even glamorous in its small way, his earnings were unpredictable. How long he stayed with this group, or why his connection with the gang was discontinued, is not known although his separation was said to have been honorable.[8] He was described as a good-natured easy mark, with little inclination for work and a total inability to get along without drinking. He was very shy, "couldn't open his mouth unless he was drunk," but his shyness kept him from getting into fights. In spite of a reputation for being "sharp" as a gambler, Sayosuke managed to lose enough money to force him to put up his wife's land as security for his gambling debts.

In the "disastrous" first year of their marriage, Shika gave birth to a girl. Since it was the year of the dog,* and since farming families often gave their daughters the names of animals whose attributes they hoped the child might develop, the little girl was named Inu, meaning "dog."

Many years later, when Inu was married and had children of her own, she showed more outward signs of affection for her father than anyone seems to have shown at any time. When Noguchi made his carefully staged homecoming trip in 1915, Inu and Sayosuke stayed together in the background while Shika toured the country with her illustrious son. On another occasion, however, Inu told a story that indicates how seldom Sayosuke was at home during the years of her early childhood. She remembered playing in front of their house one day when she was three or four years old, and a man came by and patted her on the head, saying

---

*The calendar in use at this time was a Japanese adaptation of a Chinese calendar.

"You're a good girl." Not recognizing him, Inu was frightened and ran to Shika who told her that this man was her father.[9]

Sayosuke held various jobs, such as a mailman and a porter, but there is always the implication that he was a wastrel. Although he had talent as a craftsman, his creations gained him no respect.[10] He designed many objects, some practical and some ornamental—bird cages, fishing poles, chairs and other furniture, carved figures of animals—and made them of wood that he selected in the mountains. None of these has been preserved. Only his son, Hideyo, may have been touched by his father's efforts. Hideyo would later carve his own declaration of independence in wood when he first left home, a plaque that can be seen at his birthplace today. It proclaims "I shall never set foot on this land again until I achieve my objective."

For Sayosuke had indeed served the purpose for which Shika tolerated all of his shortcomings. Two and a half years after Inu's birth, Shika finally had her son, whom she named Seisaku.* Having fulfilled this vital need for the perpetuation of her family, Shika was now willing to undertake full responsibility for that family which consisted of the two children, her drunken husband when he was there, and her ailing mother. Shika's continued tolerance of her husband's inadequacy was explained in the community by her understanding of the role of the Japanese wife and of her fate as a Buddhist. The mothers of famous Japanese are generally portrayed as long-suffering and self-sacrificing, and no doubt this was often the case.

Soon after the birth of his son, Sayosuke left to take employment in the house of the headman of a neighboring village, where he was when his son's tragic accident occurred. If Sayosuke ever participated in any way, or if he was ever consulted, in the planning for his son's future, that fact has not been recorded.

*He changed the name himself to Hideyo.

2

Childhood

> Whosoever hath anything fixed in his person that doth induce con-
> tempt, hath also a perpetual spur in himself to rescue and deliver
> himself from scorn.
>
> —Sir Francis Bacon, "Of Deformity"[1]

AS THE BOY SEISAKU GREW THROUGH INFANCY AND CHILDHOOD TO SCHOOL
age, he was strong and healthy enough in body. His life was dominated,
however, by poverty, the constant state of his mother's hard work, and the
ridicule to which he was subjected by other children because of his conspicuously mutilated hand. The children threw stones and taunted him
with shouts of *tembō,* an epithet that means "arm like a wooden stick."
Okumura writes of Seisaku's fantasies of separating the fingers himself with
a knife, of his running to hide from his persecutors behind the memorial
stones in the cemetery, and of his finally learning to fight. The stories told
to Okumura by the villagers are all that is known of those early years;
Noguchi never mentioned his early life in his letters nor have his friends
left any record of his preschool years.

Shika was obsessed with the fear that her crippled son could never
become a successful farmer, and her restless groping with the ever-present
poverty drove her to take on a job as a porter during the winter when no
work could be done on the farm. There was always cargo of some kind to
be carried to and from Wakamatsu. Cargo was then still carried by human
beings, although this was considered to be work for men, and only strong
and rough men at that. When Seisaku was five years old, he went in the
morning to see his mother off on her route and met her when she returned

at night. By this time he was aware of Sayosuke's dissipation of money on *sake* and had seen his father take money that his mother earned through such labor. Shika had ample reason to feel hemmed in and frustrated during the long winter months when she could not work in the fields, and she had so little with which to provide, or even to plan, for her family. But she was a strong-minded woman, far from accepting a passive attitude toward life. How could she continue to allow her husband to take even the money she earned?

In their patient, gentle, passive compliance, Japanese wives appear to both Occidentals and Orientals alike as an infinitely desirable ideal of femininity. In Shika, however, there was a striking similarity to the long-suffering "wives of alcoholics" as they have been described in medical reports and in literature of Western culture.[2,3] At least one Japanese psychiatrist finds this classic picture valid for Japanese culture as well.[4] Wives of alcoholics in any culture can perhaps be described as aberrations who pervert the ideal of the compliant wife by carrying it to an unrealistic extreme, that very extremism representing their own hostility.

Shika, in this aspect of her life, seems to fit such a profile. Wives of poor farmers in Japan (or any country) have to be tough if they are to survive at all, and Shika had already shown herself to be as tough as the best. Why was that not enough for her? Why did she tolerate her husband's excessive impositions, depriving her children, especially the son on whom she counted so heavily? Wives of alcoholics have been found to have a special need to appear strong and to foster their husband's weakness in order to enhance their own image of strength. Often their own fathers were weak and inadequate, as Shika's father had been.

Shika was culturally conditioned to accept imposition, and it is also possible that she consciously played the role of the martyr to gain sympathy and help. Whatever her motives were in relation to her husband (and we cannot be sure how much the image of these two people has been distorted by time and the telling), her son's ambivalence was clear; deeply devoted to his mother and disposed by Japanese culture to respect filial obligations, Seisaku nevertheless came to see his mother's strength as cruel and punitive in her reaction to his early progress in school.

When Seisaku was first enrolled in school in the spring of 1883, a neighbor provided him with a brush, books, and other required equipment. However, during the first three years he was an indifferent student and this close association in school with the children who had previously teased and tortured him brought him even greater misery. In addition to the incapacity he suffered because of his hand, and the revulsion that it caused both in himself and in others, he was dirty, ragged, and unkempt. He skipped school frequently and ran off into the mountains where he sat all day just dreaming.

The competitive Shika must have demonstrated her disapproval of this behavior rather strongly. In his adult years Noguchi was careful to hide his left hand, and he invented stories to explain his deformity that varied according to the audience for whom he created the explanation. One of these stories, told to a close friend in New York, suggests how the child, Seisaku, perceived his mother's feelings about his progress in school. He told the friend that his mother had deliberately burned his hand as a punishment for his not studying, that the burning took place on more than one occasion, and that the fingers were destroyed one at a time.[5] There is no evidence that this was true, but it is likely that Noguchi was punished during his childhood by the burning of moxa on his body. This was a common practice in Japan at that time. It was not called a punishment but, rather, a cure. The anthropologist Ruth Benedict writes:

> This is the burning of a little cone of powder, the moxa, upon the child's skin. It leaves a lifelong scar. . . . A little boy of six or seven may be "cured" in this way by his mother or his grandmother. . . . It is not a punishment in the sense that "I'll spank you if you do that" is a punishment. But it hurts far worse than spanking, and the child learns that he cannot be naughty with impunity.[6]

Although Noguchi's distortion of the facts was clearly a conscious fabrication, the association of burning and punishment by his mother was also clearly a theme in his life.

Seisaku's schoolwork and behavior underwent a sudden improvement after his third year in school. Okumura, who substantiates this change by citing school records, suggests that a new teacher was responsible for the change. Helpful though this new teacher may have been, other influences must have also made some contribution to the change.

Seisaku did have one friend during his elementary school period, a boy named Daikichi Noguchi, whose family owned the adjoining farm. The family relationship is said to have existed so far back in time that it is no longer a matter of record. Chobei Noguchi was the neighbor who had furnished Seisaku with the equipment he needed to start school, and he had begun to teach Seisaku to read and write, together with his own son, Daikichi, before the boys entered school. Being the same age, the boys had undoubtedly known each other from infancy. They were tutored together in calligraphy after which they went to the temple together to begin the study of Chinese classics and English.

Seisaku progressed more rapidly in English than Daikichi. Still the boys are increasingly referred to during the early school years as friends, and Shika liked to think of them as brothers. We do not know how much Chobei Noguchi's help stimulated either Shika's interest in education or Seisaku's change in his work habits; Chobei certainly supplied a kind of support that Seisaku did not get from his own father. Daikichi was a true brother in the sense that Seisaku could compete with him successfully without losing his friendship. This must have been of inestimable value in Seisaku's learning to deal with the problem of the unfriendly children at school.

Whatever the determining factors, at the age of about nine, Seisaku seems to have developed a capacity for ignoring the jeers of the other children and to have become highly competitive in his schoolwork. This capacity, together with Shika's aggressive prodding, pointed him in the direction of a formal education to find a suitable occupation for a farm boy with a crippled hand.

In the summer of 1888, when Seisaku was almost twelve years old, Mt. Bandai erupted. Bandai is a two-peaked, active volcano whose previous eruptions had probably formed Lake Inawashiro along which Sanjōgata and Inawashiro villages are located. Although this eruption was violent and dramatic, the scorching of the earth that occurred was all on the north side of the mountains. On the southern side, where the lake and villages lay, the huge rocks that were blown out dammed two rivers, eventually forming two new lakes. The immediate effect was the alteration of the course of the two rivers, which severely curtailed the water supply for the rice fields for the rest of the season.

The fields of Sanjōgata dried up, and fights arose over the rights to such water as was coming through. A twenty-four-hour guard program, in which Seisaku participated, was set up to prevent any farmer from taking more than his share of water. While Seisaku was on guard duty one night, he saw that his own fields had less water than the adjoining fields of Daikichi's family. He yielded to the temptation and took some of the water. Daikichi saw him do it and hit him and knocked him into a mudhole. Shika scolded both boys, berating Seisaku for stealing, and lecturing Daikichi on the obligation to make allowances for a less fortunate brother.

Noguchi's biographers discuss this story with both censure and sympathy, elaborating with other instances in which Seisaku tried to devise means of getting water in a manner that would not be considered stealing. He built a dam, which was considered a clever thing for him to have done, but his reasoning that this somehow legitimized his taking water is not quite accepted in the discussions. Stealing water was as serious an offense as a farmer could commit at any time, and in these times it was clearly not to be condoned. But to Seisaku, the question was survival: should he maintain himself at the expense of his friends, or subvert his need to the good of the group? He had already suffered at the hands of the group as a vulnerable child trying to live with the continuing trauma of his crippled hand. In a recent biographical commentary on Noguchi, Hisaharu Tsukuba deals with the point by saying that morality is for those who can afford it. Shika and Seisaku had to use whatever came to hand, be it stolen water or even good friends.[7]

At the end of Seisaku's sixth year in school, examinations for graduation were carried out with a formality that suggests that this may have been the first class to have graduated. A teacher from Kōtōshōgakkō, the "higher elementary school" in Inawashiro, was asked to conduct the examination of the children. This teacher was an ex-samurai, Sakae Kobayashi, whose family had lived in Inawashiro for six generations and had observed the traditional code of the samurai as a way of life. When the Restoration eliminated the opportunity for samurai service to the local lord, Kobayashi did as some others of his class found appropriate because of their education; he became a teacher.

He had been a member of the first class to graduate from the newly created Fukushima Teachers College and ranked first in that class. Rather than trying to advance himself by going into the school system of a larger city, as his training and status would have permitted him to do, he was adhering to the old Confucian ideal of the "son staying with the father," symbolized by remaining as a teacher in his own rural community.[8]

Seisaku greatly impressed Kobayashi. When the children were called before Kobayashi, one by one, for examination, Seisaku stood before him — dirty, dressed in his usual rags, with his left hand carefully hidden. But Seisaku's academic performance so far excelled the others that the contrast prompted Kobayashi to make inquiries. Learning from the school about the Noguchi family and their struggles, and from Seisaku, himself, that he would like to continue his education but saw no possibility of doing so because of his family's lack of money, Kobayashi invited Seisaku to bring his mother to discuss the matter.

When Shika and Seisaku were admitted to Kobayashi's house, Seisaku immediately proceeded to the Buddhist altar common to every Japanese home, prostrated himself, and offered prayers before either he or his mother spoke a word.[9] Shika followed with the presentation of some freshly caught shrimp wrapped in straw. Then the discussion began. Kobayashi was greatly impressed by Shika's pathetic story and by her "formality."[10] Shika used the Aizu dialect, and in no way could she have sounded like an educated woman of an upper-class family.* But she knew how to show respect and humility in an acceptable manner. This meeting was the beginning of a relationship in which Shika and Kobayashi understood each other well. Kobayashi became the first of Noguchi's patrons, one of several who remained his patrons for the rest of his life. Both Noguchi and his

---

*The Japanese language reflects a recognition of the class structure and, fully as much as the British if not more so, the Japanese know instantly a great deal about each other simply from the manner of speech. Deference is shown through the choice of words and forms. Knowing how to speak "politely" means using a formal and deferential style in which the speaker humbles himself and shows respect for the person addressed.

mother had a talent for attracting and holding the interest and sympathy of people for whom helping others had some special kind of appeal.

As a result of the interview, Seisaku was accepted as a student at Kōtōshōgakkō without a tuition charge. It was intriguing news in Sanjōgata that Seisaku, a poor man's son, was to become a student at this school. Ordinary people thought that Shika was mad to entertain such high aspirations for her son, but grudgingly they admitted that the boy showed ability.

Seisaku began his studies at his new school in April 1889. Its standards were unusually high for such a small village, and the teachers are said to have been excellent. Many of them went on to become distinguished members of their profession. An especially unusual feature of the school was the English language course. There was a choice of two curricula, one for agricultural studies and one for commercial; Seisaku chose the agricultural. The school has no exact counterpart today. Its primary function was the training of children who were expected to become farmers or to engage in small business in the local area. The school did not prepare students for admission to a university, but was probably acceptable as background for the new teachers college or for other professional schools.

Seisaku now studied more than was required and, moreover, he reacted with a quick, peremptory defense to any notice taken of his hand. When a friendly boy asked him how his hand had been so damaged, Seisaku hit him. The school was in the center of Inawashiro village, about four miles from Sanjōgata. At first, Seisaku walked through all kinds of weather, but again Chobei Noguchi came to his aid. Chobei owned Matsushimaya, an inn in Inawashiro, where he gave Seisaku a job stoking the bath fire. This provided light and warmth for study, and Seisaku stayed there as often as he went home.

During this period Seisaku learned of Napoleon; he was particularly fascinated by Napoleon's ability to get along with little sleep and he decided to emulate this pattern. Seisaku maintained a strenuous schedule of work and study, and when Daikichi suggested that it was not healthy he would say "Look at Napoleon!" Noguchi's lifelong pattern of long hours of work with little sleep was firmly established at this time.

At home, however, things were no better and Shika suggested the adoption of a husband for Inu in order to get another worker for the farm. Inu objected, declaring that such an adopted son would expect to replace Seisaku in his position as the future head of the household. She took a job in another household to bring in some money for her mother, but the loss of Inu's help at home was not offset by her earnings. Shika then arranged to bring in the daughter of another branch of the Nihei, a girl named Otome, as a potential wife for Seisaku and to replace the work Inu had been doing. Now Seisaku refused the marriage and the girl returned to her home. Inu later married a man who was somehow persuaded to become a part of the Noguchi household and to work on the farm without adoption. The details of this marriage are not known, but this man also took to drink early in his life.

The physical conditions and attitudes that dominated the lives of the Noguchi family are most realistically described by Magoroku Ide. In his preparation for a biography of Noguchi (a project that he later abandoned), Ide went to Sanjōgata where he saw the "awful house" in which the family had lived. In an article he wrote about the project, Ide described the pill bugs scuttling across the dirt floor and the barrel dug into the ground under the sink. The men urinated directly into this barrel and all other used water, excrement, and waste were thrown into it. The contents of the barrel, after fermentation, were customarily taken out and used for fertilizer. When one visits the restored house today, one feels a wave of nausea, thinking of the smell that had emanated from this barrel.

Ide reported stories, told him by contemporaries of Seisaku, of fights between Seisaku and Inu in which they both battled to avoid inheriting the house. Inu could not face bringing an adopted husband into such squalor and tried to avoid the marriages that were offered her. In one wild fight, she and Seisaku ran out of the house both threatening suicide, and the neighbors had to subdue them physically.[11]

Seisaku did make some new friends at school in spite of using his time so intensively for study and for helping at home, but even these were friends on whom he could call for help. Two boys in particular, Akiyama and Yago, began to render financial and other material assistance to Seisaku, encouraging a dependency he did not know how to avoid. But Seisaku's most important achievement of this period came from an unconventional use of his skill with words.

Japanese children were not encouraged to express their personal feelings; rather, the exact opposite was the case. They were indoctrinated with ideals of self-control and learned early that the lot of the farmer was to "bear what cannot be borne." Nevertheless, Seisaku wrote an unusual, but undoubtedly genuine, description of his feelings about his maimed hand as a composition for a school assignment. His composition came to the attention of Kobayashi, who was so deeply moved by it that he showed it to other teachers and had it read before the whole school. As a result, the students and teachers collected a fund for Seisaku's use to consult a Western-trained physician in Wakamatsu. The fund was far from adequate to cover the costs and Kobayashi is said to have privately made up the difference, but he felt that this was more properly a community project, for its effect on the givers as well as the receivers.

Dr. Kanae Watanabe, the Wakamatsu physician, had originally come from the Aizu area. Watanabe told Seisaku that he could separate the fingers surgically and restore some independent movement and function, but explained that the end joints, which had been destroyed, would not grow again. Even this, however, would be an enormous improvement, and a date was set for the operation. Akiyama accompanied Seisaku to the hospital, and after the operation was successfully completed, he is said to have run all the way back to Sanjōgata with the happy news.

When Yasuhei Yago came to accompany Seisaku home on the day of his discharge from the hospital, they went first to a photographer to have a picture taken of themselves before walking back to Sanjōgata. This was to become a pattern in Noguchi's life: one of his first reactions to a happy event was to have his picture taken in a way that would identify and memorialize the occasion. In the first of these pictures, the kimono-clad adolescent bumpkin with the bandaged hand looks a most unlikely prospect to become a man of the world.

In the spring of 1893, at the age of sixteen, Noguchi graduated as the top student of his class. The few days he had missed from school because of the operation were the only days he had missed during his entire four years, but they were enough to exclude him from being cited for honors. A flowery petition was drawn, however, and submitted by the principal to the district government office, requesting that an exception be made in his case because of the unusual circumstances. The request was granted and honors were conferred on him.

Shortly before the end of the school term, the school had required that the students purchase a new textbook that cost three yen. Needless to say, Seisaku did not have the three yen, and when he asked his friend for help, Yasuhei Yago gave him his own copy. Yago now betrayed the fact that he was not without his own inner conflicts; he made complicated plans to replace his book by stealing the funds from his father's money box. When Yago could not find an opportunity to do so, he told his father what had happened and asked for the money, which the father gave him without hestiation. Seisaku acknowledged this by giving the father an IOU for the sum involved, a document that the senior Yago preserved and that now has considerable value among collectors of Noguchi memorabilia.[12]

With graduation and the successful operation behind him, Seisaku had substantially improved his situation; yet it was not at all clear how this improvement was to be put to practical use. He wished to enter the teaching profession, but when he made application for a position in a school in Sendai, he was rejected because it was felt that his deformed hand would prevent his giving the instruction in physical education required of all teachers.[13] He consulted with Kobayashi who suggested that he inquire whether Watanabe might accept him as a student.

Biographers commonly hold that when the operation on his hand

Yasuhei Yago and Seisaku Noguchi, 1892. Hideyo Noguchi Memorial Association

proved successful, Seisaku immediately determined to study medicine. He may have confessed this ambition to Kobayashi, planting the seed for the suggestion that he apply to Watanabe; Kobayashi, himself, was vague on this point. He was concerned about Seisaku at this time, convinced that the boy needed more guidance than Shika was able to give him. Kobayashi had no children of his own and went so far as to offer to adopt Seisaku, but Shika refused.

The suggestion to approach Watanabe delighted Seisaku, who leapt instantly from his despair to elation and left at once for Wakamatsu to seek an interview. After a long wait, Seisaku was admitted, only to be told by Watanabe that he had no need for another student. Watanabe, however, agreed to take Seisaku on as one of the helpers he employed to do casual chores. This was good enough for Seisaku. He walked back home after the interview, and by the time he arrived home his fantasies had turned him into a medical student.

3

Medical Chore Boy

WHEN SEISAKU NOGUCHI MOVED INTO HIS NEW POSITION IN WATANABE'S household, he presumed to enter a world that had completely excluded him in the past and was far from ready to make him one of its own. In the twenty-five-year period since the Restoration, Japan had adopted a constitutional form of government that outlawed class distinctions and forbade all special privileges of the samurai, including the symbolic wearing of swords. But law is one thing and custom is another. The samurai no longer paraded his swords, but class distinctions dissolved slowly. Japanese society is still far from free of these distinctions today, a hundred years later. In Wakamatsu, the back country castle-town of the Aizu lord, the old ways yielded even more slowly than in those areas where Western technology had become available.

After completing a university education in Japan, Watanabe had studied medicine at the University of California at Berkeley from which he graduated in 1887. Such an education strongly suggests that his family were people with influence, perhaps also wealth. In Wakamatsu he was an elected member of the local assembly and he owned a substantial amount of property.

His clinic, Kaiyo Iin, was housed in a two-story building, which was quite Western in its exterior design. The Watanabe family still live in this house, but a gas station has replaced the walled front-entry garden of Watanabe's day. The clop-clop of wooden *geta* (clogs) can still be heard through the sounds of cars and trucks passing on the wide paved street, a far different road from the narrow, crooked lane leading to the house in

38

the 1890s. In Noguchi's day, Kaiyo Iin housed the entire Watanabe family, including the doctor's mistress who lived in one of the rooms designed for hospitalized patients. There were treatment rooms and consultation areas as well as housing for ten or twelve students and for the many servants that were required by such an establishment.

Watanabe practiced and taught contemporary Western medicine as he had learned it in Tokyo and the United States, but the structure of the hospital-clinic-household was drawn from the medical apprentice system of pre-Restoration days.[1] Although the students served like internes, since they were graduates of a university or professional school, they were still selected largely through their family connections. The students were interested in practical experience, which academic programs did not provide, and they helped care for patients admitted to the hospital, assisted with those who came for consultation, and compounded the medicines dispensed. As they learned the students progressed to greater responsibilities, which included teaching the newer students. One man in Noguchi's time, Hasemura, was considered the best student, but because he had recently been ill with typhoid fever, he was unable to do much clinical work and was assigned entirely to teaching.

Seisaku arrived wearing a new, lined, cotton kimono, a new cotton *obi*, and *hakama*.* Shika had managed to scrape together enough money to buy the *obi* and the cloth for the kimono; Akiyama had contributed the *hakama*. Although the coarse cotton was the least expensive material available, the clothes were clean, in contrast to his previous mode of dress. These, however, were modest attributes in the environment of students with whom he was now associated.

Shortly after Seisaku's arrival, Watanabe took on another young man, Yoshida, but under somewhat different arrangements. Yoshida paid for his own clothing and personal expenses, and in addition, he supplied four baskets of rice per year in payment for his board. Aside from this difference, the two boys were more similar to each other than to the other students. Watanabe housed them together in a twenty-four-mat room (a very large room, undoubtedly planned for some other purpose) where he found them studying together as late as one or two o'clock in the morning. Like Noguchi, Yoshida was bright and ambitious and he, too, had no previous medical training.

Seisaku gradually worked his way up to the status of a part-time student. The students were given complete access to Watanabe's extensive medical library that included books in English, German, and French as well as in Japanese. Watanabe further arranged for Yoshida and Noguchi to study

---

Hakama are Japanese culottes, worn as a formal element of attire with the kimono. An *obi* is a sash, usually of firmer weave than the kimono and of complementary color and design.

with an English teacher in the town, and encouraged them to study German as well. When Akiyama, who was studying Chinese literature in Wakamatsu at this time, learned that Watanabe's father had a fine library of Chinese classics and poetry, he joined with Noguchi and Yoshida in requesting permission to study these scrolls. It was here that Noguchi made his first crude attempts to write poetry in the Chinese style, an effort that brought out his ability to involve others with his emotions and especially touched Yoshida.

Watanabe had ordered a microscope from Germany. When it arrived, all the students lined up to see bacteria, in this case spirochetes, under 1,200-power magnification. For Seisaku and Yoshida, this was an exciting new experience. When Noguchi's turn came, he spent such a long time looking that the group grew impatient, prompting one student to remark that Noguchi would never be able to use a microscope anyway because of his hand. Yoshida was so sickened by this mean-spirited remark and the pain it caused Noguchi that he left the line without taking his turn. A tiny incident, but one which suggests that Noguchi may have been treated with some contempt by the more fortunate students. Although nothing is known about Noguchi's student life at the Watanabe clinic beyond his friendship with Yoshida, it seems likely that his experiences added to the resentments that already tortured him but that fed his ambition.

A student would not have made such a demeaning remark to a person who was his social equal or superior; the concept of social status, the sense of inferior and superior, was so deeply embedded in the Japanese culture that it was expressed almost automatically. It is still reflected in the Japanese language as one learns it in childhood, in the concept of obligations and relationships between children and parents, between siblings, between students and teachers, between employees and employers. It is a basic element of all relationships between people in Japanese society. All the young people of the lower classes of that time would have to face it every day in every relationship of their lives as a contradiction between the dreams that were stimulated by the promised changes of the new laws and the reality as they experienced it. Noguchi had been especially sensitized to it by the experiences of his early childhood, and his reactions were not only intense but became a permanent part of his adult personality.

In the latter part of 1894, when Japan became embroiled in a war with China and Watanabe was called for military duty, Noguchi's life changed abruptly. Watanabe closed the clinic and found places elsewhere for all of the students except Noguchi and Hasemura. Noguchi did not have the "background" to be placed and Hasemura's health still prevented his carrying a full load. Watanabe assigned some responsibilities to Noguchi so that he could remain in the household; Hasemura, on the other hand, remained as a guest. Menial though Noguchi's interim job was, Hasemura was angered that responsibility had been given to the lowest member of the group whereas he, the senior student, had none. Having nothing to do, he occupied himself by stirring up trouble for Noguchi.

Noguchi's job was that of a petty steward and errand boy. The agent who managed Watanbe's real estate gave Noguchi the money to make ordinary purchases for the day-to-day needs of the household. As might be expected, there were two interests to be served, those of Watanabe's wife and family, and those of his mistress. Noguchi was caught between them. He extended primary consideration to the wife, and Hasemura seized on this to encourage the mistress to complain that she was being denied proper respect. Noguchi thought he should withdraw altogether, but when he sought Kobayashi's advice, Noguchi was told that the experience was good training for him and that he should try harder.

When Watanabe returned in the spring of 1896, he was pleased with the manner in which Noguchi had managed his household. Noguchi's report, said to have been written in English but no longer available, accounted for all receipts and disbursements down to two sen for a postage stamp. Watanabe was even more favorably impressed with Noguchi's considerable progress in studying the medical literature. With this encouragement, Noguchi immediately began a campaign to persuade Watanabe that he was ready to take the first half of the national examination for a license to practice medicine.

Noguchi had found other opportunities to improve himself during the year and a half of Watanabe's absence. A group of French Catholic missionaries had established a church and school in Wakamatsu, offering Seisaku an opportunity to study French and to make acquaintances among the membership. He was baptized, joined the church, and participated at least to the extent of singing in the choir. He learned languages easily and rapidly throughout his life. He not only enjoyed cultivating this ability but often used it to ingratiate himself with strangers in Japan as well as in countries throughout the world.

Noguchi continued his writing of Chinese poetry, notably to make overtures to a girl who was not as easily impressed as Yoshida. He had seen this girl passing on the street and had fallen in love with her on sight. Her name, he learned, was Yoneko Yamanouchi. She was a friend of Watanabe's sister, a member of the Catholic church, and a student at the missionary school. She was sixteen years old and lived with her mother who was a doctor's widow. Seisaku identified her situation with his own since her mother, too, was having a financial struggle, and he felt all the more strongly drawn to her because of it.

Seisaku's first approach to Yoneko was a love letter, written in the form of a Chinese poem. He signed it with the name of Yoneko's girl friend,

Machiko, but added the initials "S. N." in Roman letters. Unable to understand the letter and certain that it had not come from Machiko, Yoneko showed it to her mother who took it to the principal of the school who could not understand it either. Seisaku continued to write letters and poems to Yoneko, throwing them through the open door of her house, until he was finally identified as the writer and was ordered to stop. He never managed to have a word with Yoneko in Wakamatsu, but they were to meet again elsewhere.

During the summer following Watanabe's return, a young dentist from Tokyo, Morinosuke Chiwaki, conducted a dental clinic in Wakamatsu, As the only dental facility in the town, the clinic was popular and busy. Because Chiwaki, a man of some social standing, had come with a formal introduction to Watanabe, he spent many of his evenings visiting at Kaiyo Iin. On such occasions, Noguchi displayed his usual flair for attracting attention, looking somewhat unkempt but conspicuously reading foreign books. As Noguchi expected, Chiwaki asked him how he had learned the foreign languages. This seeming paradox often gained this kind of attention for Noguchi; he was well aware of it and cultivated the technique. He told Chiwaki that he had learned the languages from the missionaries and managed to add that he intended to go to Tokyo to study medicine. Chiwaki replied that he would be pleased to have Noguchi call on him in Tokyo, the very reply that Noguchi had sought since he was busily making plans for his next step.

By the end of the summer Noguchi had Watanabe's endorsement to take the medical examination. The first section covered the basic sciences, such as anatomy and physiology, and the second half dealt with clinical applications. It was not at all uncommon for students to fail each part several times; they expected to remain in Tokyo for an indefinite period to study and repeat the examinations until they passed or gave up.* The examinations were given only in Tokyo and to live there one needed money.

Confident that he would pass the basic science half of the examination on his first try, Noguchi planned to find a job to supplement the funds from Inawashiro while he prepared to take the clinical section. He wrote of his plans to Kobayashi, who gave his consent and promised what little financial help he was able to give. With that, Seisaku went home to bid farewell to his family. When he departed from Kaiyo Iin, Watanabe gave him a letter to Chiwaki and ten yen. Okumura later refers to a promise of ten yen a month with the vague implication that this promise had been made by Watanabe. The money was never forthcoming, nor is it at all

---

*Graduates of university schools of medicine were not required to take the national examinations.

certain that the promise had been made, either by Watanabe or by anyone else, but Noguchi seems to have expected it.

He had learned that Yoneko also was going to Tokyo to enter nurses' training, and before he left Wakamatsu he tried to see her. But he saw only her mother who gave no encouragement to the relationship. On the day Noguchi returned to Sanjōgata, his father came home sober. His father indicated that he felt deeply the implications of his son's going off to study in the loneliness of a strange, huge city with no help from his family, and he tried to express his understanding and sympathy for the undertaking. Seisaku could not avoid seeing that his own neglect of his duty to his parents appeared even worse in the face of his father's negligence, but when he saw how miserable his father felt when he was sober, Seisaku could do nothing but accept the situation as it was.

Noguchi visited his friend Daikichi, who gave him three yen. Then, with Oshika,* he called upon Kobayashi, who gave him some advice and ten yen. Apparently, Noguchi sought no help from his friend Yago, the boy who had given Seisaku his own book at school and then tried to steal the money from his father. Noguchi was to call upon Yago so freely for funds during the next few years that only some special consideration would have prevented his doing so at this time, yet Okumura makes no mention of the point.

Evidence from Noguchi's own hand at a later date suggests that he wished to conceal something about the time he spent at Wakamatsu. In a curriculum vitae submitted immediately after his arrival in the United States in 1900, Noguchi altered a number of facts relating to his educational background, apparently to make it more impressive, at least to Western academicians. He changed the dates of his stay at Kaiyo Iin to support an altered statement of his medical training, and to remove from the record any evidence that he had worked at Kaiyo Iin during Watanabe's absence.[2] On the surface, Noguchi's concern appears to relate only to his status in the household.

In a letter written to Kobayashi from Philadelphia in 1901, he remarked, "I feel more at ease since I no longer face such conditions as I did in Tokyo and Wakamatsu."[3] Okumura wrote freely about Noguchi's deplorable behavior in Tokyo, yet he allowed this equating of Tokyo and Wakamatsu to slip by without comment, as though encouraging his readers to search between the lines for explanations that, for some reason, could not be stated. Whatever the Wakamatsu experience had been in terms of Noguchi's social acceptance or other problems, it had been genuinely productive in preparing him for the medical examination and in

---

*At this point in his biography, Okumura begins adding the Honorary "O" to Shika's name and I follow his precedent.

expanding his education in foreign languages and Chinese literature. But he felt that he could gain nothing by remaining longer; he was impatient to move on.

The railroad had not reached the Aizu-Wakamatsu area yet.[4] The nearest train for Tokyo ran through Koriyama, which was accessible only by twenty-five miles of footpaths through the mountains. In September 1896, Noguchi set out alone on this trail with his twenty-three yen, the letter to Chiwaki, and almost nothing else.

4

Up to Tokyo

THERE IS NO WAY OF KNOWING WHETHER NOGUCHI HAD EVER BEEN TO Koriyama, or had ever seen a train, when he first set off for the capital. But nothing could have prepared him for Tokyo, which was then, as it is today, the largest city in the world. Nevertheless, he quickly found lodgings, took the first half of the medical examination early in October, and before the end of that month he was notified that he had passed. Recording neither surprise nor elation at this remarkable achievement, he began to look for some means of supporting himself, but his momentum died out as he failed to find a job. Money that he had expected to receive from home did not arrive, and by the date of the emperor's birthday, November 3, he was penniless and apprehensive.

Noguchi had hoped to find a job before calling on Chiwaki at the Takayama Dental College. Instead, he made his first visit in a spirit of defeat in spite of having passed the examination. Pouring out his story, he dramatically appealed for a job as Chiwaki's *rikisha** man. Chiwaki pointed out that such grinding labor would leave him too exhausted to study, and he arranged with the college cook for Noguchi to earn his meals and a place to sleep by helping with the janitorial chores of the establishment.

Chiwaki was one of those men who welcome the challenge of fighting for a new idea or for a courageous underdog. Only a few years older than Noguchi, but far more sophisticated, he shared Noguchi's unorthodox

---

*A passenger vehicle drawn by a man.

approach to traditional restrictions. Chiwaki is spoken of as a gambler who took chances on people as well as in business. His choice of a career in dentistry was typical of his interests; dentistry was not yet a recognized profession in Japan, but was regarded more on the level of the barber's trade.

Chiwaki's family background gives some clue to the origins of his rebellious nature.[1] He was born into a family named Kato. His Kato grandparents were prosperous people who had a number of daughters but no son. Following the established custom, the grandfather adopted a son for marriage to his oldest daughter and the couple had a son whom grandfather Kato accepted as his heir. Not long afterward, however, grandmother Kato gave birth to a son. The grandfather broke the contract under which his daughter's husband had been adopted, and designated his own newly born son as his heir. Feeling some obligation to the young couple, he searched for records of a family that had died out because of a lack of heirs, and found the Chiwaki family. He then arranged for his grandson to take up the succession and ultimate responsibility for this household; thus the disinherited child became Morinosuke Chiwaki. The child had all the benefits of the usual heir to the head of the household — the Chiwaki — but was still recognized as the grandson of Kato. For practical purposes this meant that Morinosuke was exempt from military service, but that he had the social and financial backing of his grandfather, at least through the period of his education. Chiwaki graduated from Meiji Gakuin and Keio University, both of which were, and still are, institutions of high social and academic prestige and very costly for students.

Chiwaki had taken his degree in English literature, and after graduation, had worked on a newspaper and taught English and English literature before he discovered his interest in dentistry. When he later became head of his own dental clinic and school, he was affectionately called *oyabun*, the boss, but without the implications of criminal activity so often associated with the Japanese *oyabun* and their gangs. Rather, his leadership conveyed the quality of the family style group so popular in Japan; a group of men who engaged in various enterprises and were bound together by the old-style family loyalties to their leader and to each other.

As *oyabun*, Chiwaki used his connections with influential people to further the interests of his group and its members. In the social climate of Tokyo, such sponsorship provided identity for the group members in addition to meeting their prerequisites for acceptance in any field of endeavor from the business and academic communities to the marriage market. For Noguchi, identification with such a group was, in effect, the discovery of his second patron.

Noguchi now decided to undertake further study of the German language. After the Restoration, the Japanese had adopted the German model for their universities, largely because Germany held the foremost position in the world in science and medicine. Noguchi found a German lady who gave instruction for one yen a month, and he asked Chiwaki for the money. Since Chiwaki earned only four yen a month, he turned to the director of the school, Dr. Takayama, who reluctantly raised Chiwaki's salary to seven yen a month. Of this, Chiwaki passed on two yen a month to Noguchi, who worked at his German with great intensity for three months. By the end of February 1897, Noguchi announced that he now knew German and discontinued the lessons.

When Noguchi studied, he was able to concentrate to the exclusion of almost everything else. What he excluded now, however, was some of the work he was supposed to do for the dental college. Some of the chores were difficult for him, for example, cleaning the soot from lamps, which were hard to hold with his crippled hand. But he also neglected to ring the morning bell, which required little enough skill, his preoccupation with his own concerns being somewhat greater than his sense of obligation.

Noguchi planned to take the second half of the examination in the fall. By May of that year he had concluded that he would be wise to enroll in Saiseigakusha, a privately owned medical school that had been established by Dr. Tai Hasegawa to meet the demand for Western-trained physicians after the Restoration. Heavily oriented toward cramming for the national examinations, the teaching at the school was done by young assistants from Tokyo University Medical School who earned fifty sen an hour and an honorarium of one yen a month for lecturing in their spare time. They read their lectures from prepared texts; when one text was finished, the teacher started again at the beginning and read his assigned series over again. The students could come in at any time and stay until they had learned enough to pass the examination, if they could afford to remain that long.[2]

The lectures covered causes, symptoms, and treatment of the disorders a physician was most likely to encounter in practice, with some attention to the special problems of the military. Without laboratories or clinics, the students practiced what techniques they could on each other. Noguchi had trouble learning to hear chest sounds because his deformed hand prevented his performing adequate chest percussion. Again Chiwaki stepped in. This time he arranged for a second operation on Noguchi's hand, which not only improved the appearance and function of the hand but gave Noguchi an opportunity to learn chest examination technique from the surgeon who performed the operation.

At the lectures Noguchi assumed an arrogant manner, often standing at the back of the crowded lecture room, either taking no notes or ostentatiously taking them in English. Since his budget rarely permitted him to visit a bath house, he was as dirty as he was arrogant and consequently he was generally disliked by the other students.

One student added to Noguchi's pleasure only by her presence: when he arrived at Saiseigakusha he found Yoneko Yamanouchi already enrolled. She had finished her training as a nurse and was now working part-time while studying to take the examinations to become a physician. She and Noguchi finally achieved a speaking acquaintance; he helped her with explanations of the lectures and even gave her a skull as a gift. But she still had little liking for him.

He did develop a friendship with a cousin of Yoneko's named Kikuchi, a medical student who worked as a policeman to support himself while he was studying medicine. Proper and serious minded, Kikuchi nevertheless became Noguchi's companion for evenings of relaxation that neither could afford. Their destination, like that of generations of medical students, was the forbidden pleasures of Yoshiwara, known as the entertainment center of Tokyo. It had *sake*, good food, and theater, notably Kabuki, but its chief attraction was the prostitutes who were allowed to ply their trade in this section of the city without too much interference. Even in the less expensive establishments, most students managed to spend more money than they could afford. Noguchi, who never had been able to control his finances, now began accumulating serious debts as well as a reputation as a dissolute character. Chiwaki heard many complaints about him during this time and received much advice about the fruitlessness of investing money in such a person. But he said nothing.

Noguchi continued a desultory attendance at Saiseigakusha until October 1897, when, barely a year after his first sight of Tokyo, he joined the students taking the second half of the national examination. Of eighty students in this group, he was one of the four who passed. He had loaned his stethoscope to a friend shortly before the examination and failed to get it back in time but, unabashed, he borrowed one from the examining professor.

Noguchi's success in the examination justified Chiwaki's support, to say nothing of his own faith in himself. But again he was at loose ends. He had learned that his uncouth appearance, as well as the sight of his crippled hand, repelled patients, which discouraged him from entering private practice. Furthermore, setting up a private medical practice required money, which he lacked. Even though he was sensitive about his appearance, Noguchi seemed unable to improve it. He appealed again to Chiwaki, who arranged for Noguchi to teach pathology and pharmacology to the dental students at Takayama. Noguchi returned to live at the college, enjoying the stir he created, now a teacher where he had been a janitor only a year before. But his salary hardly provided a living and he kept up a restless prodding of everyone around him.

Under continous pressure from Noguchi, Chiwaki was able to use his connections to have Noguchi appointed as an assistant at Juntendo Hospital. As with everything else, at first Noguchi was delighted. He was at the bottom level in status with duties as a medical librarian; he was to survey all new cases admitted to the hospital and summarize any unusual or interesting findings, together with reports from the literature, in a hospital bulletin that he was to edit. His earnings were two yen a month and meals. Soon everyone knew of him, but in a short time he had this job well organized and again became restless, pestering Chiwaki to get him a raise in pay. Noguchi was constantly borrowing money from anyone who would lend it, and his salary was gone on the day he was paid.

On one occasion he asked Chiwaki for a loan, saying he needed the money for a haircut. Chiwaki took out his purse and emptied its contents for Noguchi with the comment that a student always needs unexplained money. Noguchi got the point and responded with a flowery letter of gratitude for this and all previous help, giving the letter to his roommate, Ishizuka, to deliver to Chiwaki. Ishizuka was an apprentice-student at Takayama Dental School, and a friend as sensitive to Noguchi's moods of despondency as Noguchi's roommate at Kaiyo Iin, Yoshida, had been. Ishizuka later described the sadness with which Noguchi handed this letter to him.

Another such incident with Chiwaki arose from Noguchi's unsightly appearance as he followed the other doctors around the hospital. In the course of a conversation with one of these doctors, Tawara, the subject of Noguchi's inability to buy clothes came up and Tawara offered to give Noguchi his "old Western clothes." Shortly thereafter word of this offer reached Chiwaki who became furious. Okumura paraphrases the message Chiwaki wished to convey in a letter to Noguchi: "Is it that important for a doctor to be well dressed? Do you think it is respectful to visit a comparative stranger and ask for favors the first thing? If you want anything, come to me!"[3]

That night Noguchi could not sleep. Ishizuka was aware of his struggling to compose a letter to Chiwaki, which he again asked Ishizuka to deliver in the morning. Ishizuka watched Chiwaki's response as he read the letter. Chiwaki again became very angry, crumpled up the letter, and threw it in the trash from which Ishizuka retrieved it. Relating the episode to Noguchi's biographer, Okumura, Ishizuka emphasized that Chiwaki was not seeking apologies; he wanted Noguchi to show some sign of maturity in his behavior. Noguchi's letter, dated December 18, is groveling and even incoherent at some points. He was in an obvious panic at the thought of having offended Chiwaki so deeply that he might lose his support. Although he rationalized and justified what he had done, Noguchi knew that his behavior had been a source of embarrassment to

Chiwaki. But he confessed his anguish, "If only I had not been held in such contempt!" He made a feeble attempt to project the blame onto other persons and on to his circumstances, but he knew "it is a trashy thing to do—to list excuses."[4]

The holiday time can be a sad time for lonely people; Noguchi spent the last two weeks of 1897 brooding over his situation, particularly in the light of his near rupture with Chiwaki. It is the custom among Japanese to make every effort to pay all debts before the new year: even the last hours of the old year may be spent rushing to debtors and creditors in order to start the new year bright and clean. Noguchi could do none of this, but he did make a decision. He spent this New Year's Eve, the closing hours of 1897, writing to his old friend Yago.

Noting that this was the first letter he had written to him since coming to Tokyo, he thanked Yago for his enormous help in the past and for having visited Oshika in Sanjōgata. After describing how well his work had gone, but how hard his life had been—"At times I was shamed in public and left in tears and resentment"—Noguchi came to the point.

> At this moment I am in urgent need of Western clothes. People treat me with contempt because I don't have them. I may be treated as an equal by my friends, but I cannot present myself in society in such rags. It is a matter of great concern, especially in a place like Tokyo. I would be happy if I could have 50 yen. . . . I look forward to hearing from you. As my vacation lasts only until the fifth or sixth of January, please send your reply before that time.

He closed with appropriate greetings and the request that Yago keep the details of his financial condition confidential.[5]

Okumura writes that Yago had difficulty in raising the fifty yen, but he finally sent the money on March 2, 1898. When it arrived, Noguchi met Kikuchi at the hospital, whether by chance or by plan, and they disappeared together every night until the money was gone. Piecing together the scattered fragments of letters that Noguchi wrote to Yago during the first four months of 1898, some quoted by Eckstein and some by Okumura, it appears that Noguchi had ordered clothes on the expectation of money from Yago and was stalling his tailor for payment.[6] There is no record of a letter acknowledging receipt of the money Yago sent on March 2. Barely three weeks later, however, in a letter dated March 20, Noguchi blandly wrote Yago: "At present I am in financial difficulty. I hate to ask you this, but will you please raise twenty yen for me? I need this money badly by the end of the month. If you can't raise it I will have to think of some other way, so please write me immediately. As this is an urgent matter, I won't mind receiving half the amount at a time. . . ."[7] And on April 21, he wrote expressing his joy at having heard from Yago, a letter that caused him to be "much revived and almost crazy with rejoicing and peace," feeling as though a question of life or death had been resolved.[8]

Noguchi's expressions of appreciation to Yago in his New Year's Eve letter, his very reluctance to write that letter, suggest that Yago must have done much more for him in the past than give him the textbook for which Noguchi paid with the IOU. Noguchi's biographers have had to rely on Kobayashi's cooperation in making letters and other material available. Since Kobayashi frankly censored portions of letters from Noguchi by inking them out, it is probable that he showed certain letters from this period to Okumura and certain others to Eckstein. His motive seems to have been related to his wish to create an image of Noguchi to be used as a model for Japanese children, an effort to which he devoted much of his later life. In the light of these circumstances, it seems even more likely that something had occurred during Noguchi's stay in Wakamatsu that had caused him to appeal to Yago for help, that this was his reason for not calling on Yago when he left for Tokyo, and that Kobayashi chose to withhold the story.

About seventeen years later, in the fall of 1913, Noguchi discovered that he had a cardiovascular condition that was caused by an old, inadequately treated syphilitic infection.[9] There is no way of knowing just when he suffered the primary infection, or even when the cardiovascular condition itself might have been diagnosed if he had been having regular physical examinations. Syphilis may remain in a latent phase for a period of time that varies from a few years to more than twenty years. It is reasonable to speculate that Noguchi might have acquired the infection as early as 1893, probably no later than 1908. There is no point in pursuing this matter other than to note that it could have been the problem that required some very special help from Yago, and received some very special censorship by Kobayashi, while Noguchi was in Wakamatsu. We do not know that this is true: noting it as a possibility is far from presenting evidence. Noguchi told his physician in New York that he had gotten the syphilis from a "Noguchnik," which tells us only that he had picked up some Yiddish slang by the time he discussed the matter with the New York physician who was also a personal friend.[10]

Watanabe never became Noguchi's patron in the sense that either Kobayashi or Chiwaki did, even though the job at Kaiyo Iin did give Noguchi his start in medicine. Watanabe dealt with Noguchi in a straight-forward manner, and he would not have tolerated impropriety. But improprieties may have occurred at Wakamatsu, and Yago may have paid the bills for them. Such improprieties would have been consistent with Noguchi's childish, irresponsible behavior and his disorganized mode of life in Tokyo. Still, it is clear that Noguchi managed to achieve a fair amount of serious work in Tokyo even as he had in Wakamatsu. There was a focus to his ambition that kept him on the track of constructive progress in his self-education program even while his social qualities remained at an infantile level.

In the early part of 1898, at the same time that his money-clothes problem was so acute, Noguchi was translating a German textbook in pathology that he had studied at Kaiyo Iin.* He came to some material in bacteriology that he did not understand and consulted his director at Juntendo, Dr. Sugano. Noguchi's questions led to a general discussion of bacteriology, and he expressed an interest in entering this field rather than private practice.

In the light of his poor experience with patients and his own recognition of his social deficiencies, Noguchi may have thought of this possibility on his own. But his conversation with Sugano heightened his interest in bacteriology as a field in which he could avoid a confrontation with his personal problems. Sugano told him that he should become an assistant at Kitasato Institute for Infectious Diseases, which was devoted mainly to research, but he added that this institute accepted no one but graduates of Tokyo University.

Such precedents never stopped Noguchi. He immediately pressed Sugano to speak to Dr. Kitasato on his behalf. Sugano promised to do so, and Noguchi hurried off to ask for Chiwaki's help as well. Chiwaki called upon his friend Dr. Motojiro Kawakami; they all wrote letters, and Noguchi received the appointment. Again he was a lowly assistant, again his initial duties were in the library. But his salary was thirteen yen a month. He was only twenty-one years old, yet he had been granted a real, although probationary, professional status with a highly respected group. In a land where such affiliations are the key to a man's whole future, anyone but Noguchi would have regarded this as the final achievement of the goals toward which he had devoted so much of himself and asked so much of his benefactors for so many years.

---

*Clemens von Kahlden, *Technik der histologischen untersuchung pathologische-anatomischer präparate für studirende und ärtze.*

5

At the Kitasato Institute

Everything is beautiful when it is still in a dream state, but when it
becomes a reality it is no longer so interesting to me.
— Noguchi to Chiwaki, July 19, 1907

NOGUCHI'S BRIEF AND REBELLIOUS CAREER AT THE KITASATO INSTITUTE
for Infectious Diseases was the first concrete manifestation of his limited
ability to gain satisfaction from work itself. At each step he seemed driven
to start reaching for the next, unable to stop his whirlwind pace long
enough to gain any but the most superficial benefits from his progress.

The institute reflected the will, the personality, the scientific ambitions,
and the patriotic devotion of its founder, Dr. Shibasaburō Kitasato. Born
in one of the southern domains to a samurai family whose lord had
supported a school for the study of Western medicine for a century before
the Restoration, Kitasato was a liberal in his attitude toward the development of science in Japan but a conservative in his adherence to the culture
of the past. Arrogant, domineering, highly disciplined, and hard-working,
his original inclinations had been toward a career in the military, and he
never lost his interest in horsemanship, kendō, and other military arts.

Kitasato graduated from Tokyo University College of Medicine in 1884,
by which time his interest in bacteriology had been so greatly stimulated by
the discovery of the cholera bacteria by Koch in 1883 that he accepted an
appointment in the Bureau of Public Health rather than a more
remunerative position that had been offered to him. For a year and a half,
he cultivated relations with local physicians and public health officers to

53

the point that "one may appropriately date the origins of what many later called the 'Kitasato faction'. . . from this period."[1] He further assisted in Japan's first laboratory for bacteriologic research, which was created in 1885, and he published his first paper, on the growth of the cholera bacillus in pure culture, in that year.

In 1886 the Bureau of Public Health sent him to Germany where he spent six years, working mainly with Robert Koch in Berlin. While there Kitasato established his reputation through the first successful growth of the tetanus bacillus in pure culture. Of even greater future influence, he worked with von Behring in the development of the first diphtheria antitoxin, for which von Behring was awarded the Nobel Prize in 1901. After Kitasato returned to Japan, he became restless with the limitations of work in the Bureau of Health. He established his own institute in 1893 as a private laboratory. Within a year it was taken over by the Home Ministry, which expanded its functions to include teaching, epidemiologic work, and the manufacture of immune serums and protective vaccines, and even developed a hospital. The Bureau of Public Health also was under the jurisdiction of the Home Ministry and it is said that Kitasato not only ran the institute according to his own precepts but dominated the Bureau of Public Health as well.[2]

Kitasato further distinguished himself by identifying (in 1894) the bacillus that causes bubonic plague. Other members of his staff were to make equally notable contributions: Kiyoshi Shiga discovered the type of dysentery bacillus that bears his name (Shigella) in 1898, and Sahachiro Hata, working with Ehrlich in Germany in 1909, discovered the famous Compound 606, arsphenamine, the most effective of the arsenic derivatives, and the only reliable treatment for syphilis for several decades.

When Noguchi arrived at the institute in April 1898, he paid his respects to Kitasato and was received with courtesy. As institute librarian, Noguchi made a good impression with his knowledge of English, rudimentary though it was; the staff members knew only German, and were pleased to consult him with problems in reading English. However, it was clear that Noguchi would have to start from the beginning to learn bacteriology, and it was clear to everyone but Noguchi that this would be a long, slow process. Since he was a staff member rather than a student, Noguchi received no formal instruction.[3] He had to learn through experiments of his own devising, and although others may have helped him, he had to organize his working habits to accord with the mores of the institute. For this Noguchi was totally unsuited and unprepared.

For his first experiments, Noguchi asked for three guinea pigs. That request was denied with the indignant explanation that even the head of a department could expect to have no more than five guinea pigs a year; such a request from one in his lowly position was nothing short of presumptuous. Perhaps also feeling his way, Noguchi spoke to Kitasato of his hope that he would be sent to Germany to study, and learned that there were others ahead of him on the list for this privilege and it would be at least six years before his turn came. This was a slow pace for Noguchi; controlling his behavior enough to ingratiate himself for six years was more than he could manage.

Noguchi's doubts about adapting himself to such an elite environment were overdetermined by his rage at the whole class system of Japan. Tokyo University graduates, in particular, were the very symbol of all that he hated most bitterly. Hata, who arrived at the institute very soon after Noguchi, was not a Tokyo University man; indeed, he had no university background at all, so it appears that Sugano's information that the institute took only Tokyo University graduates was incorrect. Nevertheless, the spirit of the elite was overwhelming. As soon as Noguchi arrived in such a group, he wanted to be on top. The higher the level of his associates, the more he wanted to defeat them; the more elitist they were in their attitudes, the more he wanted to surpass them.

In spite of the fact that both Shiga and Hata were living in Tokyo at the time when Okumura was writing his biography of Noguchi, Okumura reports little of the actual work that Noguchi did at the institute. Later events made it clear that Noguchi achieved a level of competence equal to students who took the institute's three-month course in bacteriologic technique, but nothing is known of the process.

Barely two months after Noguchi started work at the institute, he took a week off to go home. He visited Yago at Inawashiro, and as they drank together, Noguchi talked of his determination to become a great physician. He was sure he could do it if only he had the necessary clothes and financial backing. By the time Noguchi returned to Tokyo, only a week later, Yago had arranged to supply him with a hat, a watch, an Inverness coat, and various relatively expensive items of Japanese-style clothing. Back in Tokyo, Noguchi put his plans and ambitions in writing for the first time, setting them all out in a letter to Yago dated June 13, 1898:

Dear Brother Yasuhei:
During my visit home, I am afraid that I must have bothered you with my selfish ways. Nevertheless, my feelings are intense; I do not wish to be treated as an equal only to my schoolmates and looked down upon as a cowherd. . . . I ask you to believe in my sincerity and lend a helping hand. No matter how rough the waves of my life become, I intend to achieve what I pledged to you. Please do not regard me as just another medical student and give up on me. Although it may sound presumptuous, I pray with all my heart that you will have faith in me. The following are my plans:
From July through the end of October 1898, I plan to study

microscopy at Dr. Kitasato's Institute for Infectious Diseases at Shiba. This will cost ¥130, but I plan to make do with about ¥50. I will probably find a way to earn part of the money myself and to take this course with only a small amount of extra money.

In November 1898, I plan to leave for the United States of America where I will stay in the home of an American Professor while obtaining an M.D. degree. I then intend to go into practice for a while to earn some money (just enough for tuition). I will need ¥300 for my trip, but I can make do with a little less. (However, if funds are short I will have great difficulties and may have to work so hard that I will imperil my health.) I plan to stay in the United States for a year or a year and a half.

Sometime in 1900, I plan to leave America for Germany where I will really pursue the study of medicine. My mind is set to earn an M.D. degree in Germany also which will require my staying for two or three years during which time I will support myself.

In 1902 I plan a trip to France where I will tour around and visit famous scholars and universities. I plan to get an M.D. degree in France and stay there a while. I would like to purchase medical equipment there and return to Japan early in 1903. I will think about plans for the following time after I have returned. I list here only the important features of my plans. When I return I will be only twenty-seven or twenty-eight years of age, still young and energetic. It won't be too late for me to act as freely as I like. Medical students cannot graduate before reaching their thirties—it is rare that one graduates at twenty-eight. I will be poles apart from the new graduates of the Japanese universities, like a cloud and a glob of mud. I am not boasting—I am just revealing my plans to you, my kind brother. This is nothing like casual talk between strangers; I have told you everything, without reserve. I hope and pray that you will give it serious thought.

I cannot explain all my plans and intentions now. I will tell you more of them when I succeed at least in going abroad. If I were to explain things to you now, it would lose all fragrance, like the bud of a flower forcefully opened before its time. They say "Better leave it unsaid" so I will. Please do not tell my plans to unrelated people.

He signed the letter "A child of the lake and willow." Then, noting that a flower which falls before it blooms has not fully lived, he wrote a poem. Using the allusion to a legendary Chinese tale of a person who would rather be a piece of jade smashed to bits than a common roof tile left whole, Noguchi implied that his confidence in his own ability justified his ambitious plans. He signed the letter again, "The wild child, Noguchi, born of the lake and willow." Then he wrote: "P.S. I have a special request. I am embarrassed by being deeply in debt and am in desperate need of money. Please send me twenty bills. . . .Seisaku."[4]

The poetic expressions he used in his signatures are reproduced here in literal translation of idiomatic expressions drawn originally from Chinese. "The wild child" means an undisciplined person with supreme confidence in his own ability and views but caring little whether the world agrees with him or thinks he is somewhat mad. "A child (born) of the lake and willow" is most nearly analogous to Wordsworth's "Child of Nature" and "the wild child" to the French expression *enfant terrible*.

Noguchi's reluctance to speak of what he might do after his return from foreign study is interpreted by Japanese people to mean that he planned to become physician to the emperor. This is something one did not put into words (better leave it unsaid). But in spite of his delicacy in expressing this ambition and the poetry that followed, Noguchi seemed to find nothing jarring in the slang and content of the postscript. We have no record of Yago's response.

Just at this time Chiwaki left on an extended business trip to China. Ishizuka was to take charge of the dental college in his absence, but was also to be ready to join Chiwaki in China if he should be needed. This would leave Noguchi without the two people in Tokyo on whom he was most dependent. He certainly had formed no such relationship with Kitasato or anyone else at the institute. Perhaps his learning of these plans had stimulated his trip to Inawashiro to present his petition to Yago; these developments, added to his low estimate of his future as a staff member of the Kitasato Institute, must have contributed to Noguchi's decision that he had gone as far as he could in Japan and had to get out of that country at all costs.

Shortly after Chiwaki's departure, Noguchi was notified that Kobayashi's wife was ill and he was granted leave to go home and help care for her. He stayed in Inawashiro at least a month, keeping in touch with Tokyo through letters to Ishizuka. Although Kitasato wrote him "If you don't come back to Tokyo soon, someone will take your position," it was several weeks thereafter before Noguchi returned.[5] During his time at home he found time for several significant matters. He read a current novel that dealt with the character and fate of a young medical student named Seisaku Nonoguchi, who led exactly the same kind of dissipated life as Noguchi had been living in Tokyo and who ended in failure and disgrace. Noguchi became so upset by the striking coincidence that he consulted Kobayashi about changing his name. Kobayashi agreed and offered the name Hideyo; *Hide* was used frequently in the Kobayashi family and in this context would be considered a gift; it means "successful, brilliant"; *yo* means "the world." Westerners who have learned of this name change have perhaps come a little too readily to the conclusion that it originated solely from Noguchi's idealization of himself.[6]

The author of this novel, Tsubouchi, stated later that he had written about medical students from what he had seen of them when he was a student at Tokyo University in the Department of Literature. It does seem that Noguchi was not the only one who fell into this kind of trap in the big city. It was a time when many young men from remote parts of the country were coming to Tokyo to study. Often they did not know what they intended to study, or what they would do with their knowledge when they finished their studies. The opening of the country by the Restoration was felt as a breath of freedom among young men whose ancestors had not moved out of a small area for generations.[7],[8]

Yago made no such move for himself, but he derived a vicarious pleasure from his support of Noguchi's exploits. The Inverness coat (and probably many of the other items of clothing) had been made for his own wedding. Yago's son later reported that Noguchi had pawned the coat, and that the Yago family had sent him the money to get it out of pawn, but Noguchi had neither redeemed it nor returned the money. Still Yago stood by him.[9]

Noguchi also found time during Mrs. Kobayashi's illness to work on the translation of the von Kahlden book that he had started earlier in the year and that had led to his appointment to the Kitasato Institute. He finished the translation sometime after his return to Tokyo in September, but the publishers to whom he submitted it wanted a preface written by some person whose name was well known in the field. Noguchi asked Kitasato to make this contribution. Kitasato was reluctant, but finally agreed that if Noguchi would write the preface he would permit his name to be signed to it.

Noguchi returned to his Takayama lodgings from his discussion with Kitasato and literally forced Ishizuka to lend him five yen, saying that this preface was a tremendously important thing for him and he needed the proper atmosphere for writing it. "I need certain stimuli and have to write it while viewing the full moon reflected in Shinagawa Bay."[10] There were *geisha* houses in Shinagawa, and there are legends about a prince who wrote great poetry with his head in the lap of a beautiful *geisha*. Noguchi disappeared with Ishizuka's money and returned the next day with his preface.

Shortly after his return, Yoshida, Noguchi's friend from the Wakamatsu days, put in an appearance. Noguchi was delighed to see him and wanted to entertain him, but had none of Ishizuka's five yen left. Taking Yoshida with him, Noguchi set off for Kanda, the area of book shops and publishers. He left Yoshida standing outside the Handaya Publishing House for a full hour while he negotiated the sale of his translation for thirty-five yen and came out with the money. Noguchi said he should have had 100 yen, but since he needed the money at once he had to settle for the lesser amount. He and Yoshida proceeded to a shop where they spent several hours talking, eating, and drinking, after which they parted and never saw each other again.

When his friend Akiyama, who also was in Tokyo, fell ill, Noguchi arranged for his admission to Juntendo. The illness was diagnosed as typhoid so Noguchi had him transferred to Kitasato Hospital where Akiyama made a good recovery. Throughout his friend's illness and convalescence, Noguchi was most attentive. Noguchi needed friends, and he gave them as much as he could. But in November, when Chiwaki sent for Ishizuka to join him in China, Noguchi was really alone. We know little of what he did to get himself through the winter, but spring brought a new development.

In view of actual events, Noguchi's announcement to Yago that he planned to live in the house of an American professor in the United States was the kind of demanding prophecy that he so often was able to build into a reality. There was at least one book in circulation in Japan at this time telling of the experience of Japanese students who went to the United States and were taken into the homes of Americans who, since they were all rich, befriended the would-be scholars and supported their studies. Such stories attracted many young Japanese men to the United States.[11] Noguchi added the flourish of the American's being a professor, and there is no reason to believe that he had any more substantial basis than this when he outlined his plan to Yago. In April 1899, however, the means of implementing it seemed to drop into Noguchi's lap.

Members of a medical commission from the United States stopped in Tokyo on their way to the Philippine Islands to investigate the outbreak of various infectious diseases among the American troops stationed there. Dr. Simon Flexner, who at that time was an assistant to Dr. William Welch, dean of John Hopkins University Medical School, was a member of that commission. When Flexner arrived in Tokyo, he requested permission to visit Kitasato's institute, and Kitasato chose Noguchi, because of his knowledge of English, to call upon Flexner at his hotel to arrange the visit. Flexner wrote of this first meeting with Noguchi in his diary:

> At his first call, Nogouchi [*sic*] spoke to me of coming to America. He wanted if possible to come there for a few years to work in bacteriology. I suggested that he speak about the matter to Professor Kitasato and if he approved his going and would recommend him, that I would write to Philadelphia in the hope of obtaining some sort of appointment in the University.[12]

In a later version of this meeting, Flexner emphatically denied that he had given any encouragement to Noguchi's request. Flexner explained that he had known, when he met Noguchi in Tokyo, that he would not be returning to Johns Hopkins but was about to transfer to the University of Pennsylvania.[13]

Okumura's version of this meeting adds that Kitasato invited the visitors to dine with him and other staff members of the institute. Noguchi, because of his lowly status, was excluded from the dinner table and was required to eat in a room by himself in spite of the fact that he was to conduct the group in sightseeing the next day. Noguchi never forgot this. Years later, he still spoke of it with great bitterness.[14]

Whatever the details of the visit, Noguchi did write to Flexner in Manila. His English has a character all its own, but there is no doubt about its meaning.

> I have spoken him [Kitasato] plainly of your condinal agreement. Then Professor Kitasato approbated my hope, and let me free whether to go abroad or stay more longer, but he adviced that if Professor Flexner will do favor of recommendation it is more advantageous to go abroad the more early. Here I have mailed off his letter for you, and I wish your generous disposition. My hope is to study state-medicine, and these which are most excellent in U.S. above the world. . . . About the most important factor or the sustenance I feel very immense grieve as I am quite independent at present; perhaps you know very well about it. Please kindly let me have any deservable position in America. I never dount before any severest service through incapacity. . . .[15]

He had talked with Kitasato, who had agreed that Noguchi should go to the United States if he could arrange to do so, the sooner the better. In saying that he was "independent," Noguchi clearly meant that he had no means of support. It is possible, but not likely, that Flexner misunderstood his meaning from his incorrect use of that one word. Noguchi wanted to escape from Japan any way he could. Years later Flexner acknowledged privately in his diary that he understood this, although at the same time he was publicly encouraging the belief that his initial response had been very casual. Although Flexner kept this letter from Noguchi in his files, together with a second letter that Noguchi wrote a year later, there is no record of his own replies, nor is there a copy of the letter from Kitasato to which Noguchi referred.[16]

Although this chance meeting with a well-placed American professor fit Noguchi's plan as though he had ordered it, he still had to solve the problem of funds for his travel, and it would be over a year and a half before he could embark on his journey.

6

Physician in Yokohama and China

Teach thy necessity to reason thus;
There is no virtue like necessity.
Think not the king did banish thee,
But thou the king.

Shakespeare — *King Richard II*, Act 1, Scene 3

THE AMERICAN COMMISSION WENT ON ITS WAY, AND NOGUCHI RETURNED to his restless routine. A few of his days were brightened by an invitation to call on Viscount Chutoko Ishiguro, one of the earliest of the modern Japanese physicians. Having heard that Noguchi had served as guide and interpreter for Dr. Flexner's visit, Ishiguro wanted to know what impression Tokyo had made on the Americans and Noguchi's answers to his questions were precise. Noting Noguchi's Tōhoku accent, Ishiguro inquired about his background. Shortly after this meeting Ishiguro wrote Watanabe: "Your student, Noguchi, is brilliant and will have a great future."[1] Noguchi, as always, was pleased to have attracted the favorable attention of an important man, but still the compliments were of no help in solving his immediate problem.

In his daily activities at the institute, Noguchi had inevitably gotten into trouble. He had loaned several library books to his friend Shigematsu, books that were considered government property to be used only by members of the institute staff. When the time came for an inventory Noguchi learned that the books, along with many of Shigematsu's own

61

possessions, had been sold. Noguchi tried to buy replacement books but since the loaned library books were technical books, mostly foreign, replacements were not readily available. Noguchi had to report the loss to Kitasato, who is said to have restrained himself, but since Kitasato's temper was notorious that seems doubtful. For the time being, however, he let Noguchi return to work with only a severe reprimand.

Japan had recently set up its first program of harbor inspection with four quarantine stations, the main one at Yokohama. Kitasato had been active in this program, especially in the selection of staff physicians, and he now came up with the idea of appointing Noguchi as assistant inspector at Yokohama. Okumura wrote that Kitasato "may have" used this opportunity to get rid of Noguchi, and that he also "may have" felt that such episodes as the disappearance of the library books would not have happened if Noguchi had not been so deeply mired in his financial problems. ("May" usually means "positively is" in Japanese usage.) Noguchi started on his new job at once. As quarantine officer, his salary is variously reported to have been thirty-five or forty-five yen a month. Since room and board could be had for only six yen a month in Yokohama, this was a substantial increase in his income and provided Noguchi with an opportunity to begin to resolve his financial problems.

Noguchi was happy and he worked hard at the new job, taking special pleasure in exploring the foreign ships and talking with ship's officers from many countries. In the early part of September 1899, he found a very ill Chinese man hidden in a storage room of the ship *Amerikamaru.* Noguchi thought that the man had plague; other doctors disagreed, but Noguchi took a blood sample and found in it the plague bacillus. He reported his findings, and protective measures against further spread of the plague throughout the ship and into the local population were taken. This, of course, was the primary objective of the whole quarantine program; the identification of such a case justified the program and demonstrated the wisdom of Kitasato's judgment in its design and administration.

Such demonstrations were important for the Japanese in those days, both for resolution of their internal differences and for their international reputation. The so-called Meiji government had begun to modernize the country only thirty years before. Already, a man trained entirely in Japan had shown the sophistication to apply scientific techniques that were relatively new even in the West. For Noguchi, it was evidence that he had not altogether wasted his time at the institute; his ability to identify the organism could only have been acquired in his study of microscopy and bacteriology while working there. Beyond that, it restored some of the face he had lost with Kitasato.

Kitasato summoned Noguchi to Tokyo for further consultation. He commended Noguchi's work, and then told him of an international commission that had been established to investigate an outbreak of plague in Newchwang, China. Japan was participating in the investigation by sending a team of physicians, and he was appointing Noguchi to this team. The pay was very good, 200 tael a month or about $130 (U.S.). In comparison, Noguchi's earnings at the institute had been about $7.50 a month, and as quarantine officer at Yokohama his monthly salary was $17.50-$22.50. Kitasato was giving Noguchi every opportunity to earn his way to study in the West even though he could not send him as representative of the institute.

As quarantine officer, Noguchi had been entitled to wear a glamorous uniform that had been a real solution to his concern about his appearance. Noguchi was proud of the uniform and the picture of him taken wearing it shows him looking pleased to the point of smugness. He hated to give up this uniform to take on the China assignment, and as a last gesture he paid a call to show the uniform off to Yoneko and to tell her of the new developments in his life. She was surprised, but quickly wished him good luck and a polite good-bye. That did not satisfy Noguchi, who pressed her for more. He wanted her to go to a photographer and be photographed with him in his uniform, and he wanted her to go to the station and see him off as he returned to Yokohama. When she refused all of his requests, Noguchi told her that the last train had gone anyway and he would have to wait and take another at daybreak. At what hour she got rid of him is not told.

The physicians going to China were each given their fare from Tokyo to Kobe and ninety-six yen for the clothing and special items they would need in Newchwang. Because of Noguchi's still unpaid debts, everyone who heard of his leaving showed up to collect what he owed them. Noguchi was thus soon penniless without having nearly paid them all and without having purchased any of the necessities for the expedition. When the other doctors left for Kobe, Noguchi was still in Tokyo trying to reconcile his financial problems. Chiwaki, who had just returned from China, sold his new wife's wedding kimono for fifteen yen, but even Noguchi could not accept all this; he returned ten yen. Chiwaki gave Noguchi a used American suitcase and some used clothes, and with these, in mid-October, in a two-yen summer suit and a hunting cap, Noguchi departed from Tokyo. The clothes he had acquired with Yago's assistance seem to have disappeared. Of course, the Inverness coat had been pawned, perhaps the delegates were not wearing Japanese clothes. . . . Noguchi left from Shinagawa station to avoid his debtors who might likely have appeared at Shinbashi to see him off, and he just made it to Kobe in time to join the group.

Now that he was really on his way out into the world, Noguchi was in no mood to be held down by the distinguished company in which he was traveling. As he wrote for an Aizu news bulletin: "The wind blew quite

Hideyo Noguchi as quarantine officer, 1899. Hideyo Noguchi Memorial Association

fiercely and all the passengers lost their appetites except myself who was accustomed to the life afloat, even though briefly."² When the other doctors felt well enough to do so, they studied Mandarin from books they had brought with them, but Noguchi, "looking at them with cold scorn," went below and started talking with the Chinese crew. When they arrived at Newchwang, he was speaking Chinese of a sort, and while his Chinese was probably not very elegant, Noguchi became the interpreter for those in the group who had been struggling to learn the aristocratic Mandarin.

With the same bravado, he volunteered to speak English, French, or German, and he became so useful as an interpreter that he was kept in the central office of the Department of Hygiene at the side of the director and at the center of operations. Since Newchwang was no place for a man with only a summer suit, he had the additional advantage of working in a heated office until he could get some warm clothing. He wrote later that Newchwang was very cold, and that the river on which it was located was almost a thousand yards wide and yet was frozen from November until April.³

He made friends with Chinese government officials and even presumed to exchange poetry with the Chinese governor of the area. The plague was over by the time the group arrived, but the Japanese doctors were encouraged to remain for a while as general practitioners. Noguchi knew little about clinical work, but he learned as he went and threw himself into the effort with enthusiasm. Even as he was enlarging his knowledge of obstetrics and of opium addiction, he was well thought of as a practitioner and nothing is mentioned of problems because of his hand.

He made a minor study of the behavior of the plague bacillus in cold weather, wrote a paper on his findings in English, and sent a copy to Flexner. Within five months he is said to have been speaking Chinese with fluency, and had acquired some knowledge of Russian. But he still was not saving any money. He never knew where it went, but his month's pay was still vanishing on the day he received it. There was a wall around the town with only one gate that was locked at night. The entertainment quarter was outside the wall and Noguchi, who often failed to return before the gate closed, had to climb back over the wall. One night he was caught and remained in jail until his identity could be verified in the morning. He still had Yoneko on his mind and wrote her several letters to which he received no reply. He had two gold rings made, one for himself and one for her, with an inscription inside hers. He sent it to her, but she still did not respond.

In May 1900, when it came time for the delegation to go home, Noguchi still had saved no money. The other members returned to Japan, but Noguchi stayed for another two months working for the Russian delegation that paid him 300 tael a month. He now turned his mind back to his original purpose, and he wrote Flexner stating firmly that he still planned to go to the United States. Noguchi acknowledged receipt of a letter from Flexner, congratulated him on his new position at the University of Pennsylvania, told of his work in China, and then stated his plans. "I never change my first desire and I shall voyage in this August surely expecting to see you at your Institute. So I beg you will find any suitable position (as scholar or student) for me. . . ." Flexner wrote on the back of this letter "Very important S. F." but with no hint as to when he made this note.[4]

This time Noguchi almost made it. After having worked for the Russians for two months, he had saved 200 tael; but there was a Japanese "friend" who knew his habits, and he managed to relieve Noguchi of his savings with what Okumura calls a "fake IOU." Again Noguchi was broke, and there was no further work in Newchwang. His assignment ended on June 28, and that night he was on board a ship bound for Japan. He had little money to show for his nine-month's work in China, but he did have a citation signed by the officers of the Newchwang Sanitary Commission commending him for "great zeal and acumen in his bacteriological work."[5]

He appears to have sought no employment when he returned to Tokyo. He stayed at Chiwaki's house and concentrated his attention on breaking down the resistance of Yoneko, and in raising money for his trip to the United States. When he called at Yoneko's house he was told that she was out. He was fairly sure that this was not true, but he could do nothing but leave in a huff. He decided to pursue his fundraising in Inawashiro. Although there is no record of Yago's reponse to Noguchi's elaborately drawn petition, now two years in the past, Noguchi considered it appropriate to explore the matter again.

He visited Kobayashi and Yago to tell them that he must have 500 yen to go abroad. As usual, Yago was intrigued but Kobayashi was concerned by Noguchi's continuing lack of self-discipline. Kobayashi pointed out to Noguchi that even if he were able to borrow the money, he would be on his own again as soon as it was used up. Kobayashi also told Noguchi that he would have to start depending on himself sometime and that he would be better off to start right now. Noguchi promised to think over this advice and told Yago that he guessed he would not need the money after all. Noguchi promised to write an article about his trip for the alumni bulletin of Kobayashi's school, from which both he and Yago had graduated, and he returned to Tokyo empty-handed.

Perhaps because he was staying at Chiwaki's house, we hear no more of Noguchi's drinking and carousing for the time being. He wrote the promised article for the school bulletin, touching lightly on the sights and the work but dwelling on his observations of the influence of Westerners in China. His view was one that was commonly taken by Japanese newspapers at that time, yet he made his points effectively. He wrote of having seen three British warships "proudly at anchor" in the harbor at Weihai while two newly built Chinese warships were "wandering outside the harbor because they had no naval port." He exhorted his readers to come forth with their best efforts on behalf of their country:

> How tragic for the Chinese ships! How tragic that the white race has been infesting their territory to such an extreme! . . . we do not have to let them ride over the Orient. Don't look upon China today as some foreign country, my dear alumni. Our interdependent neighbor is now being ruined. We must stand up, . . . and revive our country in order to prevent such invasions. We must unite and *do* it — with ploughs, with the abacus, with engines, swords and pens.[6]

Knowing that these sentiments would be pleasing to Kobayashi, had Noguchi also convinced himself that he wanted to gain a Western education only for the benefit he could bring back to his country? Yet his efforts to do something for Chiwaki as well really had such a permanent effect. He gave a well-organized series of lectures in pharmacology at the dental college. He used a French textbook, but since none of the students could read it he translated as he went along, creating in the process the whole Japanese vocabulary of dental anatomical terminology that was adopted and is still in use.

But still he had no funds for his trip to the United States. In spite of Kobayashi's advice, he tried to borrow the money but without success. Chiwaki could not give it as he had put all his resources into starting his own dental school. He did not want to put pressure on Noguchi to enter private practice, perhaps because he guessed that might cost as much as the trip abroad. Chiwaki and Noguchi both talked with everyone they could think of but to no avail. Noguchi applied for a job as a ship's doctor, and when that was refused he gave in and made an application for a license to start a private practice.

No sooner had he done this, however, than a solution to his problem appeared from an unexpected source. He and Chiwaki went on an excursion to Hakone where Chiwaki met some members of the Saitō family who were staying at the same inn. The Saitōs, Noguchi learned, lived in the Azabu section of Tokyo, which meant that they were rich. Mrs. Saitō was impressed with Noguchi and the story of his struggles and ambitions; it seems she had a niece and, like a good many Japanese ladies, she also had a fondness for trying to arrange marriages.[7]

After lengthy negotiations, which were carried on back in Tokyo between Mrs. Saitō and Chiwaki, an engagement for Noguchi was agreed upon. Chiwaki was not at all sure he thought well of this; in fact, he was quite dubious. He let himself be persuaded because of Noguchi's enthusiasm, which was for the dowry rather than for the girl, and because Chiwaki simply had no other idea where the funds might be found. The girl's family promised to give Noguchi traveling expenses to the United States; he would be gone approximately two years, and the couple would be married after his return, within five years.

Noguchi left at once to say good-by to his family and friends, but we have no idea what he told them about the source of his funds or whether he told them of his engagement.' Okumura says that he told Kobayashi nothing except that he was going to the United States. But suddenly he must have realized that his family were beyond caring. Oshika is said to have been sad, but she must have been confused as well: he had done nothing for the family and now was about to strike out in directions that meant less than nothing to her. Sayosuke was too drunk to worry about it; the grandmother who had not noticed when the little boy crawled into the fire twenty years before was hardly aware of what was going on now. Inu had all she could handle with two children, farmwork, and a husband who had now become a drinker.

At this time a scene occurred that Kobayashi was to report on later with what comes through as sticky sentimentality in translation, but in spite of its tone there seems to have been something solid behind it. Noguchi and Kobayashi are pictured praying together before the home altar when Noguchi burst into tears and told Kobayashi that he was deeply concerned about his mother; that his own thinking had all been directed toward his own ambitions and he really did not know who was going to take care of her. Kobayashi promised to see that she was taken care of, and Noguchi expressed his gratitude in profuse thanks which grew to his saying that he wished to consider the Kobayashi couple as his parents. "Let me call you father." Kobayashi agreed, and they joined in a solemn mutual pledge acknowledging their new relationship.[8]

In a final word of farewell, Kobayashi reminded Noguchi that "his three assets were his mother's love, Kannon's mercy, and his physical deformity" and urged that Noguchi never forget the importance of these influences. From what we have learned of Oshika, we can surmise that she and Kobayashi shared the wish for Noguchi to rise in the world and to demonstrate that he was a special person. Oshika had counted on her son to carry out her ambitions for the family long before the accident that had deformed his hand. Far from modifying those ambitions, she had readily adopted the plan of encouraging his education.

Whatever Kobayashi's motives may have been, he carried out his pledge to Noguchi and the family for the rest of his life. There are those who say that Kobayashi made a career out of his sponsorship of Noguchi, and no doubt it was a source of satisfaction to him. Since he had no children of his own, and since he believed deeply in the old samurai concept of family obligation, this was a natural outlet. That he was opinionated, bossy, and self-righteous is beside the point.

Mrs. Kobayashi had been raising silkworms for many years, and he

Sakae Kobayshi, Hideyo Noguchi Memorial Association.

husband had "permitted her to save her earnings for any use of her choice." At this point she gave Noguchi 200 yen "as a gift from his mother."

For Noguchi, the facts were as he stated them. He had done nothing for his family, and he now faced this realization for the first time, just as he was going even farther away for an even longer time, to a way of life that could only bring further estrangement. When he returned to Tokyo from these farewells, he must have been in a state of emotional exhaustion.

The Saitō family sent 200 yen to Chiwaki in keeping with their part of the agreement, and Noguchi went to Yokohama to arrange for his passage.

He was carrying at least 400 yen, and he had many friends in Yokohama. He arranged a dinner party in a restaurant and invited some thirty or forty people to bid them farewell. His behavior at the dinner was even more strange—he came an hour early and then mysteriously disappeared, only to return without apology, saying that he had been in a billard parlor. There is surely a good deal that we do not know about this party, but we do know that the next day he needed almost all of the funds he had collected for his trip to pay the bill.

He returned to Tokyo and told Chiwaki what had happened, and this time Chiwaki was speechless. There was no question of canceling the trip—Chiwaki's face was involved as he had been the negotiator in the arrangement. Unless Noguchi left for the United States within a month, he would be considered a swindler. Chiwaki went to a moneylender, borrowed 300 yen with which he bought a third-class ticket, and held it himself until he saw Noguchi on the ship.

On the day of his departure, Noguchi's fiancée was among those who saw him off at Shinbashi station in Tokyo. As far as anyone knows, that was the last she ever saw of him. Chiwaki went with him to Yokohama, saw him on the ship, and discovered that another friend of his, Midori Komatsu, was embarking to take up an assignment to the Japanese Embassy in Washington. Chiwaki introduced them, but since Komatsu was traveling first-class he and Noguchi could not join forces until they were at sea.

Before leaving the ship, Chiwaki had to make one last effort; he would not have felt right if he had not tried. He told Noguchi that he was like the bird pushing its young from the nest; that there was no question of Noguchi's asking further help from anyone, and he had to consider himself completely on his own. This was the time for Noguchi to demonstrate whether he was a moral imbecile or a true genius, and Chiwaki hoped he would make every effort to come back a real man.

These last few weeks had brought Noguchi closer to acknowledging the feelings behind his irrational behavior than he had ever come. He was neither a moral imbecile nor a genius; he had grandiose ambitions, he could work almost without rest, he knew he had ability, and he gloried in having his ability put to a test. But being completely on his own was exactly what he feared most. So lacking was he in any capacity for independence that it had never occurred to him in all his planning that he would truly be cut off from everything and everyone he had ever known. In contrast to the brash optimism and curiosity with which he had set off for China a year before, Noguchi now found himself terrified.

He occupied himself constantly throughout the trip with reading Shakespeare. On the third day out a great storm arose, but Noguchi continued his reading. Komatsu suggested that something more up-to-date might be appropriate, but Noguchi said the basic works were representative and he wanted to start with the best.

After twelve days they put in at Hawaii, and although third-class passengers were not permitted to go ashore, Noguchi went as Komatsu's aide. They attended a banquet given by one of the local Japanese and were taken sightseeing. After five more days of reading Shakespeare, on the eighteenth day of the voyage, December 22, 1900, they landed at San Francisco.

They checked in at the Palace Hotel where Noguchi wrote his first letter to Chiwaki. A rather formal letter, with flowery acknowledgments of all Chiwaki's kindness, Noguchi told of the details of the voyage and his plans for the train trip to Washington. His only concession to his surroundings was a notation that he was writing from his room which was on the third floor.[10]

Komatsu reported that during the five days on the train Noguchi did not look out the window even once, but kept on reading Shakespeare. Noguchi went to the dining car only once a day in order to save money, but astonished Komatsu by spending three dollars on a package of playing cards that had pictures of all the cities along the route. He said he could look at the pictures later; for the moment he had no interest in beautiful views, but only in gaining practical knowledge. If Noguchi could not let down even to glance out the window, his anxiety must truly have been enormous.

7

Philadelphia

What memories are stirred by Noguchi's premature death. All the Philadelphia period. His arrival at the university dormitories on a Sunday morning "carrying gifts." Those amazingly productive years in which he took so great a share. He helped make my career in Philadelphia even if I helped make his. He was a much better craftsman although he did not (or could not) I believe plan and think better than I. Even at the R.I. it was craftsmanship, inexhaustible patience, rather than what I should call extraordinary mental power which brought his results.[1]

DR. FLEXNER WROTE THESE WORDS IN HIS DIARY THE DAY AFTER HE learned of Noguchi's death in Africa in 1928. He spoke of the time through which he had just passed as "a nightmarish day, of course, and a haunted night." His emotional words suggest his recollection of an intuitive appraisal he made almost immediately on Noguchi's arrival on that Sunday morning in December 1900. In spite of their correspondence,[2] Flexner was astonished to have Noguchi show up, and his first reaction was to indicate firmly that he had no funds for the support of Noguchi's study. Within twenty-four hours, however, he had decided to explore the possibility of using Noguchi's plight to start some investigations with snake venoms. Years later, Flexner wrote that he had been waiting for an opportunity to get this work started, but he wondered for the rest of his life why he was so often drawn to taking a chance on Noguchi.[3]

Flexner was a meticulous man, and his astonishment at Noguchi's arrival was as much a measure of his own inability to act without precise preparation as it was an indication of their faulty communication. As he told the story over the years, Flexner glossed over their exchange of letters, playing up Noguchi's colorful but naive courage in a way that others took up and followed. Thus, an image of Noguchi, which made him more comprehensible to Westerners, was created and reinforced. Dr. Paul F. Clark, a later colleague of Noguchi's, wrote:

> Several times I have heard Dr. Flexner tell with great relish how Noguchi appeared out of the blue, laden with gifts in good oriental fashion and wreathed in smiles. Poor Noguchi's smiles were soon wiped out. How little he realized that more than a simple letter to the Professor was essential to assure him a position in a strange land.[4]

Flexner was one of a large number of children of Jewish parents who had emigrated from Germany, and his own education and position had been won through his own talent and hard work. He had graduated in 1889 from a third-rate, proprietary medical college in Louisville, Kentucky, almost as primitive as Saiseigakusha. He was, nevertheless, accepted for postgraduate study at John Hopkins University where he became a favored protégé of Dr. William Welch, dean of the Medical College and the most influential man in American medicine of that day. Flexner was the first Jew appointed to the staff of the medical school of the University of Pennsylvania and he knew well what it was like to do what Noguchi was trying to do.

Flexner also recognized the courage and persistence that had brought Noguchi so far with so little. The letter that Noguchi brought with him from Kitasato was cold and arrogant toward Flexner as well as toward Noguchi;[5] perhaps it reminded Flexner of Noguchi's having served as interpreter in Tokyo and then being sent to eat in another room when dinner was served to the American visitors. Surely, Flexner recognized the panic when he told Noguchi that the university had no funds to support his study. In his years at the Rockefeller Institute, when he was known as a tyrant, Flexner was also known to take personal interest in the problems of his staff members. A tough administrator, rigidly adhering to protocol and careful about appearances, Flexner was nevertheless capable of acts of generosity and sentimentality. He ordered the lawn sprinklers turned on for the children of staff members to play in during the hot weather, and called their homes when the children did not appear to learn whether one of them might be ill.[6]

By the time Flexner wrote his biographical sketch of Noguchi for *Science*, the emotion of his diary entries immediately following Noguchi's death had given way to precision:

> The small capital with which Noguchi started on his enterprising voyage had been all but exhausted by the expenses of the long journey.

University funds for his support there were none; inquiry among Japanese officials brought only disappointment. Hence there was one thing only to do, namely to start work and wait for something to turn up. A theme in bacteriology was chosen and work begun in the cramped quarters allotted to pathology in the old medical building. Providence was, however, not unkind, and before long a patron was found.

A short time before Noguchi's arrival, Dr. Weir Mitchell, whose contributions to the nature and action of venoms are famous, had conceived the notion of a further study along the lines of immunology which was then a fresh and advancing subject. He and the writer had discussed this undertaking and were awaiting a suitable opportunity to make a start.

The matter was presented to Noguchi, who fell in with the idea, confessing of course that he knew nothing whatever of venoms and next to nothing of the methods of immunology. Dr. Mitchell provided funds, which at the outset just sufficed for the experiments and a modest sum for Noguchi's living.

It was a period of strenuous endeavor and simple living for him, but Noguchi's long struggle with adverse conditions in Japan made it one of no great hardship.[7]

Did he really not remember his own compassionate response to Noguchi's initial panic? Had he completely forgotten that he gave Noguchi money to buy a winter coat and, when Noguchi lost the money, that he personally accompanied Noguchi to buy the coat?[8]

Noguchi's own view of his arrival and three years of work in Philadelphia appears in his letters and curriculum vitae. Noguchi wrote Chiwaki from San Francisco that he would travel to Washington with Komatsu.[9] In Washington, his curriculum vitae was certified by Komatsu, but neither Dr. Flexner nor Komatsu may have known that the facts were slightly altered to put a better face on Noguchi's education and origins. In his curriculum vitae Noguchi stated that he had graduated from Aizu Academy, which had a samurai ring to it, and further, "On May, 1894, I entered the Tokio Medical College, from which Institution I graduated after three years study." In truth he had been serving as Watanabe's steward in 1894, and had attended Saiseigakusha, the low-grade cram school, for only three months. As noted previously, he also altered the date of his starting work at Kitasato Institute by six months.[10]

Noguchi had only $23.50 left at the time of his arrival in Philadelphia. Although Flexner found inexpensive boardinghouses and rooming houses for him, the first news that the university had no funds for his study was a devastating shock to Noguchi. He wrote to Komatsu, asking him to find a job as interpreter or anything else, either at the Japanese Embassy in Washington or the Consulate in New York. Noguchi told Komatsu that it was "not profitable" for him to stay in Philadelphia because he could not understand the English spoken there. Komatsu replied that if Noguchi could not understand the Philadephia English, he could not understand the New York English either; he might be able to help Noguchi after he had graduated, but for the present he was on his own.

Noguchi said later that at one time during this period he found that he had been "unconscious" in his room for several hours. He made a drawing that Okumura interprets as a representation of big waves under shining stars, battering a skeleton, *"Hone o sarasu,"* meaning bones exposed as one suffers and gradually dies in a foreign land.[11] To Tsukuba this drawing conveys a sense of shame with overtones of suicidal thoughts.[12] Although Flexner had spoken to him about the venom work on December 31, the day after his arrival, and he started work on January 4, Noguchi felt no assurance that he would be able to maintain himself. On January 3, 1901, he wrote Chiwaki, "Dr. Flexner tells me that I can stay and study at his place until some money comes from Japan."[13]

He was still very anxious when he received Chiwaki's reply. Once again, Chiwaki had not been able to ignore Noguchi's plight, in spite of his own obligations to his new dental school. He sent Noguchi a money order for 100 yen, but again he asserted that this was the last time he could help. Noguchi started for the post office to cash the money order, but when he got there he could not find it. Although he later found the money order in his room, Noguchi had to get along without in the meanwhile.

Toward the latter part of January, Flexner was summoned to San Francisco as one of three experts to settle a dispute concerning a suspected case of plague. Before he left he outlined the work he wished Noguchi to do in his absence, spoke with Mitchell about Noguchi and the snake venom work, and arranged for other men in the laboratory to give Noguchi some help and for a secretary to look in at Noguchi's boardinghouse to see that he was getting along all right.

Noguchi's assignment was to review and summarize the literature relating to the study of snake venoms, and to undertake any experimental work necessary to familiarize himself with the techniques involved. When Flexner returned after an absence of three months, Noguchi presented him with a 250-page review of the literature, and wrote home to Japan that Flexner was so well pleased that he raised Noguchi's pay to twenty-five dollars a month.

The word *toxin*, or *poison*, and the study of toxicology were rooted in knowledge among ancient and primitive peoples that powerful poisons are present in some plants and in the venom of some snakes, spiders, and scorpions. In the professional words of a scholar we are told that these early people

 practiced a type of immunization with sublethal doses of these poisons. The first scientific approach of this type was made in 1887 by Sewall, who showed that gradually increasing sublethal doses of rattlesnake venom would protect pigeons from multiple lethal doses of the poison.

Sewall's studies stimulated Calmette's famous investigations of snake
venoms and led logically to the production of diptheria antitoxin.[14]

As we have seen, von Behring had developed the diphtheria antitoxin in
Berlin. Noguchi must have known of this work, since Kitasato had assisted
von Behring in the research. But Noguchi may not have understood the
importance of the background work with venoms before he undertook the
survey of the literature that Flexner assigned to him.

In their first paper, Flexner and Noguchi summarized the "facts and
views relating to the phenomena of hemolysis and bacteriolysis," which
they found essential to the interpretation of their studies with venom.
Thus, they left a helpful record of the level of their own understanding of
immunology, a field that was in the process of development not by
toxicologists, but by bacteriologists and chemists, into the highly
sophisticated science it is today.[5]

Flexner and Noguchi knew the following: (1) that snake venoms cause
blood corpuscles either to agglutinate or hemolyze, that animals can
produce substances which combat these effects of venom, and that a
gradual buildup of these substances in an animal confers immunity to the
venom; (2) that bacteria also can be agglutinated and dissolved like
corpuscles, that animals are capable of producing substances that can
bring about such agglutination or lysis, and that a gradual buildup of
these substances confers immunity to the specific bacteria; (3) that blood
from one animal produces agglutination or hemolysis in another.
Therefore, blood transfusions were not yet possible and immunity could
not be produced in this way. They knew that the agents that cause the
destruction of the bacteria are contained not in the blood cells but rather
in the blood serum, and the serum of immune animals came to be known
as immune serum or antiserum.

This all suggested that if they could understand how the hemolytic and
agglutinating effects were accomplished chemically, they might be able to
learn how immunity is produced in animals and from this be able to make
immune serums that would, themselves, not destroy the corpuscles of the
animal to be immunized. A study of hemolysis and bacteriolysis produced by
venoms, therefore, appeared to offer some solutions to the problems in
producing immunity to bacterial infections.

Mitchell and Noguchi were enchanted with each other from their first
meeting. Noguchi wrote Kobayashi that Mitchell was over seventy years of
age and had been a professor for forty years, and that although he was
retired he was still very sharp. Noguchi spoke of Mitchell's being a member
of the Board of Directors of the University of Pennsylvania and president of
the Philadelphia Medical Association.[16] All this was quite true. Dr
Mitchell was associated with the best of traditions in medical education in

the United States. His father had been a distinguished physician who had initiated the interest in snake venoms. The father and son had made a large collection of dried venoms, and the son had made one of the early studies on the chemical composition and physiological action of various venoms.

Mitchell also had the reputation of being able to make small sums available from an anonymous source (his own pocket) for various causes. He had had some success as a novelist in addition to his career as a practicing and research physician. He had made one trip to Japan and was planning a second trip for the coming spring.

By February 2, Noguchi was more at ease, writing to Dr. and Mrs. Chiwaki (very American to include the wife) that he was working hard with the venoms. He was reading papers in English, French, and German, and was deriving great pleasure from being able to do so. Noguchi had found that although his English was supposed to be so good, he could understand only half of what was said and had asked his landlady and her daughter to help him. "Americans are all money-worshippers, yet even though I am penniless they don't despise me and everyone is kind." He marveled at all the equipment that the university supplied and that instruments, which had to be shared by the whole laboratory at Kitasato Institute, were available for each worker here. He hoped that if he could complete his assignment well in Flexner's absence, he might get a more stable position at the university later on. He wrote of Mitchell's coming trip to Japan and asked Chiwaki to entertain him.[17]

Noguchi's apprehensiveness continued to subside in spite of his limited budget. He seems to have managed without incurring any indebtedness, but it was over a year before he could even mention the anguish he had felt during the first weeks. In a letter dated May 10, 1902, he wrote that no one should try to go abroad without adequate preparation. "As a matter of fact, I thought about death, and if you are dead there isn't another chance."[18] When he wrote Chiwaki in April, 1901, he was even cheerful:

> As the proverb says (having a baby is worse in anticipation than in fact). I am working hard on the subject Dr. Flexner kindly gave me and dreaming enjoyable dreams of the dawn of my success. All the professors and secretaries are very kind to me and I am more fortunate than students who are paying tuition. I am planning to finish the snake venom work by September and if there is no other project I will again have no means of support, but I am not worried about dying of starvation. No matter how painful, I plan to stay the expected two years. Since one paper on snake venoms certainly will be published, I will have accomplished one of my purposes in coming to the United States.[19]

Mitchell's venoms had all been dried, but live rattlesnakes soon arrived from Florida. Noguchi learned to extract their venom and work with it. He admired the skill of a black man who held the snake with a forked stick while he struggled to get the venom. He was sickened, however, by having to feed the snakes with live rabbits. He said that as long as anyone was watching, the snake just looked at the rabbit that had been put in its cage; but Noguchi would leave and come back later to find the snake asleep with a large rabbit-shaped hump in its middle. He never could bear to watch the process of ingestion.

After he had worked for a year with live venom, Noguchi gave a seminar on his work. One of the professors who was present said that Noguchi's English was such that he could not understand what Noguchi was trying to do. Noguchi was needled by this, and wrote Kobayashi, "I tried to explain it in detail but my professor laughed and said that no one would understand it anyway — it is too complex."[20] In June 1901, Noguchi wrote Chiwaki of his first extra earnings. He received twenty dollars for writing a chapter on medical practice in Japan for a book comparing the medical practices of the United States, Germany, and Japan. He bought some books and clothes with the money and proudly added, "I represented Japan in this work."[21]

Noguchi used stationery from the Palace Hotel, San Francisco, for all his letters to Japan from the time of his arrival in America until September 1901 when, presumably, his supply ran out. It was an accepted Japanese custom to take cards and stationery from a hotel or resort one had visited and use it in this way. The Palace was the outstanding hotel in San Francisco in that day, and for a traveler in Noguchi's circumstances, very luxurious and expensive; perhaps he had been Komatsu's guest. But he wrote so many, and often such lengthy letters, that he obviously had quite a supply of stationery.

Throughout his lifetime of correspondence with Japanese friends, there are many special examples of long, complicated, flowery letters in which Noguchi indulged in philosophical musings on the meaning of life, on his own objectives, and on the spiritual guidance he had been given in the past by his benefactors. Although the message Noguchi wished to convey is not always clear, being tangled in his anxious protestations on so many subjects, these letters are often signs of his disturbance and restlessness, harbingers of an impending outburst on his part. On June 28, 1901, he wrote such a letter to Kobayashi. Greeting the family with the news that all was well, he continued:

> I was happy and relieved to hear of conditions at Sanjōgata through your letters. . . . I can see that I am more at ease since I no longer face such conditions as I did in Wakamatsu and Tokyo. Although I have told you before, I would like to report my present situation.

Noguchi did so at great length, repeating his reports that he was well-treated and had plenty of money to cover his needs.

I am not a student but an honorary assistant. This assistantship is not given to any but persons who have the credentials and skill. I have gotten this position on the basis of Professor Kitasato's recommendation when my Professor in this University was visiting in Japan; and second, because of the medical books and papers which I published in the past; and third, because of my academic research in bubonic plague in China.

Then came more about how well he was being treated, the importance of his work, and the amount of money that was being spent to support his research. Then,

Alas, what a shame that many rich foreign students [Japanese] spend their money to satisfy their curiosity or to purchase a degree in order to boast when they go home! I don't feel pleased with them. I feel sorry for Japan. I wouldn't want an honorary degree if it were not offered me by the University. There is no need to show off degrees among Japanese people. If one cultivates his abilities, he won't be abandoned by the world. Even if it should abandon me, I will be satisfied if I have something in which to believe.

He had met some students from Japan who were supported by family funds or through government scholarships; presumably they were young men from upper-class, rich, and influential families who had everything he thought he lacked.

As for me, I haven't changed much and my experiences have been limited. Still I feel that the acquaintances one makes before one's project is finished are of no use and I intend to present myself in society only after I publish the results of my work. In the United States the associations of ordinary people are based on money and power, while the associations of professional people depend on a person's earnestness and ability. I think I belong to this latter group.[22]

All of this strongly suggests that the Japanese students had snubbed him. He, therefore, proclaimed that he had been chosen for special treatment because of his previous record of accomplishment, and he rejected the standards by which these students had made their way. Since Japanese classmates usually become lifelong friends, with corresponding obligations, it is particularly significant that he rejected these friendships as not being "useful." Since he later backed down on this attitude to some extent, his angry refusal to associate with his countrymen at such a time may indicate that he was even more sensitive to their slights than he would have been in Japan where there were others to turn to.

No sooner had he quieted down from this outburst than a new source of uneasiness appeared in a letter to Chiwaki, dated September 10.

The assassination of the President, which took place in Buffalo on the sixth of September, astonished the whole world. The cause of the assassination was anarchism — in other words, the extreme socialists. It is very clear that Mr. McKinley's imperialism caused dissatisfaction among people, especially those at the bottom of society in a country such as the United States which is populated by immigrants. It is only natural that poor people hate the social system in which the strong prey upon the weak. It is especially true in this country, where all people are supposed to be equal, that the envy felt by poor people is great. . . . considered from another angle, most of the anarchists are refugees . . . orphans without a country are the main source of the anarchism which has brought a destructive air to the laboring class. . . .Fortunately, my country, Japan, has maintained for many centuries a strict distinction between the nobility and the common people which is just like the relationship between parent and child so there is no danger of such an outrage. In contrast, foreigners have leaders who hold office only two or three terms, and if the leader indulges in tyranny, the rise of anarchism is understandable. . . . Of course, if some powerful country should destroy the leadership of Japan, what would become of the Japanese people! Probably they would be worse than the anarchists and would become very dangerous people. I suppose it is very fortunate that our national government does not change. . . .[23]

Pontifical though he sounded, Noguchi was groping to understand the conflicts he saw in the country where people had so much more individual freedom than in Japan. But at the same time, all conflict and rebellion became highly personalized for him. He saw the anarchists as refugees — orphans without parental controls. Bitterly though he resented the preying of the strong upon the weak, he recognized that the rage of the oppressed man can lead to "anarchy" and assassination, and he was frightened by the murderous feelings within himself. If restrictions are removed, can such people control themselves? If Japanese people (including himself) were to be released from the control of their rigid class structure, he thought they might be "worse than the anarchists and would become very dangerous people." He tried to convince himself that it was better that these constraints be maintained, and then even lightened the problem by adding a postscript to his letter saying that he had heard that the president might recover.

Only six days later, he started all over. Now he tried to reassure himself as he wrote Chiwaki:

Now I have accomplished one half of my purpose in coming to the United States. The remaining half depends partly on luck and partly on my own actions. To confess it now, the ideas which I held when I came here were such narrow ones. For the most part I wanted to win honors, but now this idea has completely disappeared and I would like to devote my life to this work; if I cannot gain a position in which I can pursue my goals, I won't mind being a simple country doctor or anything, if I can lead a peaceful life, free of the fetters of an arrogant government. . . .[24]

There are two postscripts to this letter. In the first, Noguchi expressed his fear that his support would stop at the end of the following academic year (June 1902). "After that I will probably have no connection with the University. A crisis is approaching. After a few more letters, the day will come."[25] In the second postscript, he wrote in sober tones that President McKinley had died. It appears that his long philosophical ramblings, his flowery words of gratitude to his benefactors, and his denial of ambitions and feelings of rebellion had all been attempts to retreat from a sudden new fear of the consequences of true freedom for him. If he were really free to express the feelings that were brought to the surface by his being snubbed by the upper-class students from Japan, feelings made dangerously explicit by the assassination of the president, he, too, might become an anarchist and destroy himself in the process. What Okumura perceived as a deep spiritual change in Noguchi was a new set of personal insights. The freedom to earn one's position, find one's friends among men who had similarly earned their positions, goals that had once seemed so highly desirable, now seemed also to expose the anger and fear of the man who felt oppressed and now was orphaned. He was terrified and saw his position evaporating. "A crisis is approaching . . . the day will come. . . ."

In November, however, an event took place that relieved Noguchi of all such fears. A meeting of the National Academy of Science, of which Mitchell was a member, was held in Washington. Flexner and Noguchi presented the work they had been doing, a tremendously exciting opportunity for Noguchi. He saw it as his introduction to the elite level of American science, which indeed it was, and he wrote of it in the most minute detail to Chiwaki, sending newspaper clippings and a list of requirements for membership in the academy. He even calculated the ratio of the number of members to the population of the country to show how exclusive the academy was.

After the meeting Dr. Mitchell entertained a small selected group at dinner. His guests were Dr. Welch, Dr. Flexner, a few others from the so-called prestige universities, and Noguchi. In writing home of this particularly distinguished gathering, Noguchi listed the guests with their university affiliations and titles, their ages, and the positions in which they sat relative to each other at dinner, all matters having precise significance in Japanese society. "I felt a certain strangeness to be among people who had really achieved something in their lives. I would like to add that two Japanese names were mentioned at this meeting—Dr. Kitasato and Dr. Kakichi Mitsukuri. . . ."[26]

Noguchi was euphoric, and as a young man of twenty-five who still mutilated the English language and who had arrived in the country less than a year before under such inauspicious circumstances, he had reason to be proud. During this trip to Washington, Mitchell recommended Noguchi to the Smithsonian Institution for a Bache Foundation

Fellowship, which was awarded to him. The age of fellowships and supported research had not arrived, and although the Bache fellowship carried no funds for personal stipend, it supported the research and lent it considerable prestige. This grant, in addition to the favorable impression Noguchi made through his courteous and gracious manner, and the exotic flavor with which Westerners often spiced their views of a Japanese, all combined to put him in the spotlight and he loved it.

Within himself Noguchi was forming a most grandiose impression of the importance of his work which was, at this point, barely beginning. He wrote to Chiwaki that a paper was to be published under the joint authorship of Mitchell and Noguchi, and added:

> . . . I asked Dr. Flexner to add his name; but Dr. Flexner thought that three names were too many and seemed to be quite perplexed but he didn't have the courage to strike Mitchell s name so he was caught in the middle. I couldn't do anything and kept silent. . . . My guess is that the Professor will use his name and my name on the next paper (which will be a larger work). . . .[27]

Noguchi spoke of Flexner's plans to have the paper published in a German journal and his own plans to submit it to a Japanese journal and to send reprints to Kitasato. In the same context, Noguchi spoke of having met Prince Ito, who had visited Philadelphia. Though Prince Ito was temporarily out of the frequently changing leadership of the Japanese government, he was one of the most important figures on the Japanese political stage at that time and had been since the 1880s. Noguchi would never have met such a man in Japan.

Noguchi's first paper was published in the University of Pennsylvania *Medical Bulletin* and also appeared in expanded form in the *Journal of Experimental Medicine* under the joint authorship of Flexner and Noguchi. An unusual first paragraph resolved the problem of the authors' names:

> I have long desired that the action of venoms upon blood should be further examined. I finally indicated in a series of propositions the direction I wished the inquiry to take. Starting from these the following very satisfactory study has been made by Professor Flexner and Dr. Noguchi. My own share in it, although so limited, I mention with satisfaction.
>
> S. Weir Mitchell[28]

The explorations into immunology that had begun in this way continued for seven years as the main focus of Noguchi's activity. Before moving away from the subject of venoms, he published twenty papers (some with Flexner, some alone) and wrote the chapter on venoms for a leading medical text.[29] Noguchi's major summation was a comprehensive monograph, published by the Carnegie Institution in 1909.[30] Dr. Clark writes:

> In this country the studies by Noguchi's fostering friend, Mitchell, and Mitchell and Reichert, were probably more important than those of Flexner and Noguchi. The detailed descriptions of hemolysis produced by venoms and of the specific damage to the endothelium of blood vessels resulting in edema and hemorrhage were, I believe, the more significant contributions of Flexner and Noguchi.[31]

Clark's evaluation of the work was restrained; as research it was creditable, but perhaps no more than that. The outstanding features of the work were the quantity, the circumstances under which it was performed, and its importance to both Flexner and Noguchi as it provided their own education in the concepts and techniques of the coming field of immunology. Flexner felt this as he wrote in his diary:

> It was fortunate that Noguchi quickly learned the usual immunological technique. My own mastery was far less than was required and the multiplicity of things to do would have made my learning it slow as well as difficult. He quickly learned it and was able to use it with celerity and accuracy.[32]

For Noguchi it was, above all, a training period in the language and life of a completely foreign laboratory. He had made an impressive beginning not only in the work but in establishing a personal reputation. He was known to the most influential and able members of the scientific community as hard working, meticulous, persistent, competent, and—not the least important—utterly charming. The anxieties that were expressed in Noguchi's letters to Japan, and which were never to be entirely resolved, were not even suspected by his American associates.

8

The Lonely Foreigner

I can't help shedding tears in thinking of my past. . . .I realize that I am
still in the land of the enemy and will try to spend my time only on work.
—Noguchi to Chiwaki[1]

WHEN NOGUCHI'S EXCITEMENT OVER THE WASHINGTON MEETING
dissipated, the inevitable holiday gloom enveloped him. He had asked
Chiwaki for money for the holiday gifts he thought he had to give, and he
received the money with the usual ultimatum. In a New Year's letter to
Kobayashi, Noguchi told of the meeting in Washington in more realistic
and subdued tones. Speaking of Mitchell's offer to allow his name to appear as joint author of Noguchi's paper, he said:

> He is directly connected with my research, thus I made him coauthor of
> my thesis. If I try to monopolize the credit for one or two discoveries, it
> might be unprofitable as I might fail to find a way to continue my
> research smoothly. I accepted his offer gladly because it is also very ad-
> vantageous to have my name appear together with the name of such a
> great authority for my debut work.[2]

In spite of the obvious truth of these statements, their tone is callous and
opportunistic, especially since it seems likely from his previous reports that
he may have had a little nudging from Flexner to reach even these cold-
blooded conclusions.

Noguchi had asked Kobayashi to send certain items of Japanese clothing
that he intended to give Flexner as a Christmas gift. The articles arrived,

84

and as he thanked Kobayashi, Noguchi went over a dozen details of their beauty. He said they were "truly representative of genuine Japan" and that Flexner was pleased but had taken the opportunity to suggest that Noguchi not wear Japanese clothes during the daytime. Noguchi added that "probably a picture of [Flexner] in Japanese clothing would be difficult."[3]

He then wrote of a New Year's tea party at his lodgings to which eight Japanese came: "They are older than I, either graduates of various universities or scholarship students."[4] The group decided to hold regular monthly meetings and elected Noguchi as their president. "In a foreign country," he noted, "everything depends on one's ability and capability. Being a scholarship student chosen by the government doesn't mean much—this is an entirely different world."[5] He had come to a level of accommodation with the Japanese, but he was still gloomy. He thought he might have gotten all the benefit he could from his work in the United States and was worried that he still did not see how he could get the money to go to Germany. He spoke of borrowing and suggested that he might ask Mr. Saitō, the uncle of his fiancée, to lend him the money; he wondered what Kobayashi's opinion of this would be (as if he did not know.) He thought of stopping his research and taking a job to earn the money. In an attempt to reconcile his desires with his conscience, Noguchi tried moralizing:

> No ordinary rich man would even consider lending me money. Besides, borrowing is not suitable for my project and also is too risky. I have strictly forbidden myself from borrowing from ordinary men. . . . Taking a job would certainly deprive me of the social prestige, honors, and influential friends I have acquired. . . .My engagement for marriage was with the understanding that I stay abroad for only five years, so if I stay here longer I should ask them to break the engagement at the earliest date so that I don't cause them damage. Also I should pay back the three or four hundred yen I borrowed. . . . Mr. Saitō's niece is far less well educated than I had expected. . . . Would this action hurt my morals and righteousness greatly? Is such an action unforgivable? I have no actual relationship with her. She could get married to any man just as if she had not been engaged to me. I saw her only three or four times at her uncle's house. Dr. Chiwaki wrote me recently that I might go ahead with my plan. Besides, Mr. Saitō does not seem ready to provide me with any special assistance after I return home.[6]

In spite of enjoying his work, Noguchi was insecure in his position, lonely and homesick, with no money and no *geisha* to turn to even if he had money. In his routine of laboratory work, study, walking three times a day to eat the strange food of his boardinghouse, and finally returning alone to his room, the winter must have seemed long and dark.

He had made one friendship with a young Japanese student, Nobuyoshi
Kodama, a man about whom we know little. Kodama had come to
America about a year before Noguchi; he was studying in New York, and
had come to Philadelphia for only a brief stay "to extend his research in
medicine." We are told that Noguchi and Kodama lived together for a
short time during this winter, and it appears that Kodama's leaving
Philadelphia contributed further to Noguchi's desolation.[7] Noguchi always
wrote Kodama in English, and there appears to have been a bond between
them that made it possible for Noguchi to write in a more open and personal manner than he did with any of his other correspondents. Early in
February 1902, Noguchi wrote Kodama:

> Can you imagine how deep my disappointment was when you left
> Philadelphia for New York! My loneliness is now doubled as compared
> with the days before I knew you. Last Saturday I went to Mrs. Morris's
> party for Japanese. There were two very famous ladies, and Suzuki,
> Otaki, Ogawa, and Fujioka. . . .At the Biblical meeting, that famous lady
> made an eloquent speech which usually sounds like a sermon. Otaki said
> he was interested, but I couldn't suppress my smiles. . . .[8]

It had probably been his friendship with Kodama that had made
Noguchi's accommodation to the Japanese group possible, and he wrote of
them to Kodama again in April:

> Japan Club has been growing and we have become closer to each other.
> . . .I showed your letter to them all and they wish to be remembered to
> you. . . .on the fourth Saturday we will meet at Otaki's place and make
> *koikoku*. Otaki has a keg of *miso*. Ogawa, since he is the youngest, will
> make the dish. I am very happy that I can taste the cooking of these fine
> University graduates. . . . Please write me whenever you feel homesick
> because when you are homesick, so am I. There is no love or joy!![9]

Each of Noguchi's longings appears to have stimulated the others, for
now he wrote to Kikuchi, the cousin of Yoneko, his first love in Japan.
From Kikuchi's reply, Noguchi learned that Yoneko had married and that
really depressed him. In a very sad letter Noguchi told Kikuchi that the only thing that could satisfy him now would be to surpass Yoneko's husband,
and he vowed never to fall behind this man.[10]

To his old companions at Chiwaki's, on March 2, 1902, at 2 A.M.
Noguchi wrote: "I have missed many nights sleep and many meals, but
my brain and energy are doubled from the old days (probably because I
have given up three vices—smoking, drinking, and women)."[11] Of course,
he had not given up smoking, he could not borrow the money for drinking
and women as he had in Tokyo, and his feeling of deprivation was intense
even though he tried to make light of it with such braggadocio. By March
13, the strain had become too great and he burst out with another of his
long, desperate effusions to Chiwaki. After pages of generalities and

repetitions, Noguchi started the philosophical speculation about his future:

> . . .I have been thinking that I am no good at the art of living in the Japanese sense. But my dear teacher, in this free country where class-consciousness does not exist, the Japanese talent for getting by smoothly cannot succeed and I think what works best is the spirit of spontaneity in taking one's chances. One says what he means, and receives the answers of both sympathizers and opponents. If this were Japan, people would consider it rude not to bow three polite hypocritical bows to start with, and then to express one's ideas very indirectly; without this superficial politeness people would refuse him without even considering the right or wrong of his desires.[12]

After further complaints about the people in Japan who had refused to help him, and an elaboration of the theme of the good fortune that had brought him into association with such men as Mitchell and Flexner, Noguchi announced that he had discovered that the antivenom that was currently in general use was not effective against the venom of the American rattlesnake. He proposed to produce an effective antivenom himself, adding that it would fill an important need.

> It may annoy you that I sound a little boasting, but my research on snake venom has surpassed any in this field so far, and my way of research is daring. Therefore, my nickname in this University is Seriousman (*kiwadoki yatsu*).[13]

Okumura reproduces Noguchi's letter with "Seriousman" written in English with the words *kiwadoki yatsu* following it in parentheses, suggesting that Noguchi thought, or wished to imply, that the Japanese words were a translation of the English; Japanese dictionaries and native speakers tell us otherwise. *Kiwadoki yatsu* means a "dangerous guy," an aggressive fellow who pushes the rules to the limit and beyond. Clark, who spoke with several men who worked in the laboratory with Noguchi in Philadelphia, describes the level and flavor of their mutual misunderstandings:

> His life in the laboratory was sometimes happy, sometimes painful, but always puzzling. Noguchi beamed when Yates termed him the "yellow peril." Limericks became important in his learning colloquial English. He would cock his massive head with its mop of curly black hair on top of his small body (about five feet one inch in height), look up at the speaker with a blank expression on his face and a serious question in his eyes. Then when he began to realize that the apparently serious statement was just an American joke, his eyes would light up and a delighted broad grin would break out whether he understood the joke or not. That he was learning was apparent when one day he put up a sign on some of his experiments
> No-touchi
> Noguchi.[14]

The seriousman, the *kiwadoki yatsu*, the yellow peril—all make an agonizing conglomeration of distortions; yet it would appear that the joking use of the term *yellow peril* in this instance came closer to *kiwadoki yatsu* than Noguchi's own version of his nickname.

Continuing the letter in which this distortion occurred, Noguchi wrote several more pages about the relative positions of important people in the university and his relationship to them, emphasizing that he was being treated as a person of importance. He brought all this to the point of his concern, which was that although his research was well supported and he was treated with respect and consideration, he was still a thirty-dollar a month assistant with no stable relationship to the university and no prospects for raising the money to go to Europe. He listed six possible courses of action, discussed each one in detail, and then wallowed in confusion. His idea for producing the rattlesnake antivenom had been brought forward as a means of raising this money, but he did not see how this could be accomplished.

> If nothing works out after my contract, there is no question but that I can go home to Japan from the U.S. with flowers. It is obvious that American medical society has great hopes for my future and would send me home with honor. From an American view, study in Germany does not change a man, and this is true. But in Japan people do not think so; they look with contempt on those who return from the United States. I know very well that no one will recognize me, no matter how proudly I leave this country. It is for this reason that I have been thinking of such unlikely means as I have listed in order to go to Germany, even sacrificing the chance to leave this country in glory. . . . But I can imagine that I would be very unhappy and in despair if this should actually happen . . . . I thought of asking Mr. Saito and if he refuses I thought of asking someone else, but this plan disappeared right away because it would be unreasonable on my part and it would be unpleasant for me even to succeed in this way. I cannot break any promises. I would rather be an unsuccessful man than an immoral man. Therefore, while I may have mentioned this to you before, I have never mentioned it to Mr. Saitō. . . Please forget about my mentioning Saitō. . . . I realize that I am possessed of a single track mind and I am easily heated and become involved. . . . but if I am not accepted by today's scholastic world in Japan upon my return, I might turn into anything—like a leopard.[15]

In truth, Noguchi understood so little of what was going on around him that when he learned of a plan to send him to the Marine Biological Station at Woods Hole, Massachusetts, for the summer, he was at a loss to know what it meant. Writing to Japan of this assignment, Noguchi said that Mitchell had arranged it because Noguchi had been so uncomfortable in the heat of Philadelphia during the previous summer. "This plan was made by Dr. Mitchell and Dr. Flexner several months ago, but I was told only last month when I was invited to a small party by Dr. Mitchell." He said that he was going to study the effect of snake venom on fish, "a kind of research which has never been done before and I will have this most interesting problem for myself."[16]

Noguchi wrote of the recent establishment of the Carnegie Institution (which he mistakenly stated belonged to the government) and of the fact that this institution would support his summer work. He was impressed by the amount of money that Carnegie had given to establish this institution, and by the amount granted for his summer work. Still unsure of what it all meant for him, Noguchi wrote:

It is easy to see the difference between Japanese and American ways if I compare myself with Japanese government students. They are given $800-$900, but spend it all for their pleasure. Only two or three out of ten publish papers. Even if they write papers, their work is on a small scale and from very little data; they collect quotations from books, intending these papers for Japanese rather than foreign readers. . . . I, on the contrary, am given just enough money for living expenses while $200-$300 is spent on my research. . . .* No matter how much work I do here, it is obvious that when I go home I won't be given a good position since they have a strange thing called "academic cliques." I am determined that I will combine forces with great scholars of the world and make the Japanese give me a good position, even though they do so only out of regard for these scholars. My professor is already preparing to make an effort to have me made a professor in Japan. . . .[17]

Noguchi seemed to be pressed in so many directions at once that he could not make a decision which would last more than a few minutes. He went to Woods Hole in June and, for a change, was pleased to find several other Japanese working there for the summer. During his hikes and picnics with his fellow countrymen Noguchi took advantage of their company to discuss his ideas for making rattlesnake antivenom. In a letter to Kodama written late in July, he mentioned patents that had been granted to Kodama; perhaps Kodama, as well as his new friends at Woods Hole, encouraged his plan to produce the antivenom as a solution to his financial problems.[18] Late in August, Noguchi wrote Kobayashi as though his plans had already been accepted:

My project will now enter the treatment phase . . . so it will be more logical for my career. I will keep several goats and render them immune to the venom of American snakes. (This is very difficult and no one has succeeded in doing it up to this time.) . . . Even an American scholar of high status cannot obtain support for such experiments . . . the amount of money is enormous. Ah—I am so pleased with myself! No government student, or student on his own, can do what I will be doing. If I can achieve what I plan to do and invent the cure for American snake venom . . . and if we are lucky, I can raise the money to have you travel

---

*The Japanese stipends were $800-900 a year; the American research funds were $200-300 a month.

all over the world. . . . Ah—one poor child born in a stable at Sanjōgata, who was found by you in the old castle town, is now standing in
the middle of the active academic world of the U.S.[19]

Noguchi seemed to think that he could use his research funds at the
university for the animals and equipment he would need to develop and
promote the antivenom, and it did not occur to him that Flexner might
veto this venture into the business world. It was Noguchi's first approach to
a subject that he would raise later many times. His constant and frequently
ill-considered plans for making immune serums and vaccines not only
created many of his later problems but they also became a part of the
Noguchi legend at the Rockefeller Institute. Corner reports one small story
of Noguchi's complaining to a colleague, "He won't buy me a goat! Why
won't Dr. Flexner buy me a goat?"[20]

In these early days, the production of immune serums and vaccines was
still a part of research for which pharmaceutical houses were not staffed or
equipped. Nevertheless, it was thought inappropriate for academic institutions to undertake manufacturing and sales activities. Ehrlich and von
Behring had gotten into highly competitive battles, with errors and ill feeling on both sides, in just such activities in Germany.[21] Noguchi was certainly aware that this was an important activity at Kitasato Institute in
Japan even though he may not have known at this time that it was a focus
for the political wrangle between Kitasato and the staff of Tokyo University.[22] Although Noguchi raised this subject with every disease organism
with which he worked throughout his professional life—and there were
many—it was dropped almost as soon as it was mentioned to Flexner in the
fall.

Back in Philadelphia, Noguchi was suddenly notified that he would be
sent to Europe a year later, in the fall of 1903, for a year's study. Now even
more confused, he wrote Kobayashi pages and pages of elaborate expressions of gratitude and complicated declarations of his moral objectives,
even reverting to the use of the Tōhoku dialect in his elation:

> I cannot help shedding tears when I think of Sanjōgata. I can't do
> anything but ask your help. I was able to come to study in the U.S.
> because of you, but when I think of my home in the late hours of the
> night, my feelings become very strong. . . . I am to be allowed to go to
> Europe next year. (I haven't asked for this myself, but the University
> decided that I deserve it and they will support me with sufficient funds.
> No Japanese has ever been allowed to go abroad from a foreign Univer-
> sity.)[23]

It is now entirely clear that Noguchi had only the most limited
knowledge of what was happening in the circles in which he was involved.
Partly he was not told the whole story, partly he did not understand that
which he was told, and partly he made his own unique interpretations

derived from his cultural background and his egocentric ambitions. The facts are that John D. Rockefeller had initiated plans for the establishment of the Rockefeller Institute for Medical Research in New York, that Flexner had been chosen as its first director, and that Flexner had decided to take Noguchi with him as an assistant.

These plans were being developed in an atmosphere of the greatest caution and secrecy since the year 1897; although Flexner knew of his own appointment as early as June 1902, he was not free to make it known publicly until he had given proper notice to the University of Pennsylvania and until Rockefeller was ready to make a public announcement.[24] But Flexner surely knew of his appointment when he arranged for Noguchi to go to Woods Hole, and he surely had decided to take Noguchi with him by October 1902, when he told Noguchi of the plans for sending him to Europe.

Flexner may have broken his silence when Noguchi approached him about the antivenom work, as this matter abruptly disappeared from Noguchi's communications. Still assuming that he was going to Germany, Noguchi found new lodgings with a German family in order to practice speaking the language. In writing Kobayashi about it, he said he would repay the Saitō family. " . . . It's not that I can't see how poor my family in Sanjōgata is, but since I plan to repay their debts when I return to Japan, I feel it is better to pay Mr. Saitō while I am still abroad."[25] He thought he would be given a lump sum for his expenses in Europe before he left, and out of this he would repay the Saitō family. "It is obvious," he tells Kobayashi, "that they will change their attitude from that time on. I assume that the engagement will be broken."[26]

But all at once Noguchi was told of the establishment of the Rockefeller Institute, that Flexner was to be its director, and that he, Noguchi, was going to Copenhagen instead of Germany in preparation for joining Flexner at the new institute. In an explosive outburst to Kobayashi, Noguchi was elated and confused, jumping from one subject to another; even for Noguchi it was hard to believe.

. . . Dr. Flexner received a letter last week from the Professor at the University of Copenhagen which said, "We know of Dr. Noguchi's excellent work through his publications. We will welcome him wholeheartedly when he comes this fall. . . ." Even though I have been successful, it does not mean that I can obtain the same honor and glory when I return to Japan. Good things should never be expected. . . . I do not resent that I do not have a wife. I believe it would be a basis for my misfortune to create formal relatives. It's all right if I cannot have a satisfactory child. There are plenty of unfortunate children who need to be cared for. I would like to dedicate my life to repayment of the great favors I have received. If I grow old in Sanjōgata . . . there is nothing for me anyway and there is no question that I am still Seisaku [son] of a poor farmer. I will never forget my own background even though I have succeeded somewhat in life with your help . . . . No one can imagine how happy I am.

He then blandly reported that he had written to the Saitō family asking for a loan of 4,000-5,000 yen, and telling them that he intended to stay in the West for seven or eight years. After some interval, he had heard from the Saitōs refusing his terms and urging him to return to Japan and meet his obligations. Noguchi said he intended to refrain from correspondence with them for a while, and he expected them to break the engagement. " . . . but deep in my heart I feel pity for the girl, when I think how she must feel about the whole thing. . . . When I think of it, it's not her fault and I can't just discard her. . . . Anyway, I will return the 300 yen this fall before I leave for Europe. When I return this money I presume the Saitō family will ask me to break the engagement."[27]

Returning to the subject of his great good fortune, Noguchi did not come right out and tell Kobayashi that he had been offered an appointment to the new institute. Perhaps, in the Japanese way, he knew that Kobayashi would understand that this was the case. Enclosing a picture of himself in academic robes, he explained: "I have been given the right to wear these at this University which considers me equal to the holders of doctor's degrees."[28]

Among the available letters from this period, there is no letter written to Chiwaki between March 14, 1902, and August 3, 1903, the period during which the upheaval around Noguchi's plans for going to Europe was at its peak. It is not believable that he did not write to Chiwaki during this time. He usually wrote more freely to Chiwaki than to Kobayashi, but we are left to speculate on the contents of missing letters. When he wrote to Chiwaki in August, Noguchi was more outspoken than he had been with Kobayashi:

I am going to be a member of the Rockefeller Institute from next year on, and when I return I will be appointed head of a department. . . . I plan to stay in the U.S. five years or more. . . .the future looks better here than in Japan.[29]

Dr. Flexner took up his appointment to the new institute on July 1, 1903. Since he was occupied with planning physical facilties and recruiting staff, he could not provide any work for Noguchi. During the month of July, Noguchi spent two weeks at Schwenksville, a little resort village in the Pennsylvania mountains where he wrote home at length of the contrast he saw between farming life in America and Japan. Foreign though it was, Noguchi seemed to be more at home than he was in the city.[30] He followed his description of the village by a return to the subject of his engagement. He had learned that the funds would be disbursed to him in quarterly installments during the year, and so he could not pay the Saitōs before leaving. Tortured by indecision, Noguchi wailed to Kobayashi, "The only, the most unbearable thing for me is the Saitō affair. No matter how I think about it, I just cannot break the engagement after having her wait for three years and especially without any fault on her part.[31]

Even though he came to no conclusion, his indecision had none of the arrogant quality of his periods of elation or frustration. Gone were his proclamations of concern for his family, his gratitude toward his benefactors, his raging at the academic cliques and social customs of Japan. It would appear that these had all been attempts by Noguchi to cope with the anger and feeling of isolation that had become overwhelming when he faced an uncertain future with little understanding of his surroundings. There had been so many influences upon his life that derived from, or contributed to, his isolation: his maimed hand, his poverty, his anticipation of rejection by Japanese, his defensive competition with Japanese students or with Yoneko's husband, his confusion in the cultural atmosphere of the West, his dependence on patrons who were involved in affairs he did not understand, and his need for friends, for a woman in his life. He had kept trying to escape from his isolation through concentration on work and the pursuit of status. But when he had faced uncertainty or the possibility of failure, he could only rant and rave.

Flexner planned to embark on a tour of European laboratories at about the same time as Noguchi was to leave for Copenhagen. Though Flexner's trip was partly on behalf of the new institute, it was also his honeymoon. Noguchi was invited to the wedding, which was to take place during the latter part of September, and his sailing date was chosen with the wedding date in mind. He would surely have felt it an honor to have been asked to attend. Mrs. Flexner's family were prominent members of the Quaker community of Baltimore. Her father was a physician who served on the governing boards of several educational institutions, her mother was a leader and devoted worker in charitable, temperance, and religious activities, and her sister had become president of Bryn Mawr College.[32]

However Noguchi spent his time during this summer with no work to do, he continued to struggle with his personal problems as though he wished to start on this new phase of his life with a clean slate. For the first time in his life, his assured income was adequate to provide him with a reasonable standard of living and his professional future had prospects of a stability he had never known. Yet, six weeks prior to his sailing, he was terribly excited but still unsure:

> I am no longer the representative of Japanese medicine, but now represent American medicine. Courage is boiling in my mind. Thinking of my old days in Inawashiro makes me wonder about life's changes. . . . Whether to abandon [my engagement] or not confuses me. I will write and report everything to Chiwaki of my feelings and ask his opinion.

The damage done by the engagement so far should be repaid by me. I feel a moral obligation to do this. Even if I should support her throughout her life, it wouldn't be very much and I am not afraid at all. I will ask your opinion again later. I am ending this letter today because I am so confused.[33]

Optimistic yet apprehensive, knowing little of what lay ahead of him in Copenhagen or in the more remote future in New York, he fell back to fashioning dreams as he embarked for Europe—his ultimate objective since the days of his servitude in Wakamatsu.

9

At the Little Institute in Copenhagen

I am not going to a small country because of lack of money but because I wish to study under the most important scholar in the world. That is exactly why I deliberately chose Denmark.
　　　　　　　　　　　　　　　　　　　—Noguchi to Kobayashi[1]

BARELY HOURS BEFORE NOGUCHI WROTE THESE WORDS, HE KNEW ALMOST nothing of Denmark, of its new institute, or of its young director, Thorvald Madsen. Dr. Madsen had visited Flexner at the University of Pennsylvania during the summer of 1901, when Noguchi was still a novice, and they probably did not meet. Even if Noguchi had seen Madsen inspecting the laboratories, he would have regarded this youthful visitor as being far too young to be of any importance. It was beyond Noguchi's experience that Madsen could have been appointed director of the laboratories of a new *Seruminstitut* although he was only thirty-one years old.

In later years Madsen loved to repeat a conversation he had with Noguchi shortly after the latter's arrival in Copenhagen. Having heard of Japanese veneration for age, Madsen asked Noguchi, "How old do you think I am?" Noguchi quickly responded, "About sixty-five years." After they became better acquainted Noguchi confessed that he had wanted to show his respect by saying "one hundred years" but Madsen looked so very young that even Noguchi could not extend courtesy that far.[2]

In spite of his apprehensions about Denmark or Madsen, Noguchi is said to have found his year in Denmark to be the happiest in his life. Evelyn Tilden, who became Noguchi's assistant at the Rockefeller Institute in later years, reports that he said as much to her and that she attributed his happiness to his respect and affection for Madsen.[3] In reminiscing about Noguchi after his death, Madsen said that Noguchi had told him he "never had it as good as at our little *Seruminstitut,* and he often longed to be back in his happy youth."[4] The conclusion seems to be supported by Noguchi's mood when he returned to the United States at the end of that year.

If it is valid to single out this year of Noguchi's life in this way, one must go beyond these statements for substantiating evidence that this unique condition was more than a surface impression. If this was, in fact, Noguchi's happiest year, that would imply that his optimism bloomed in this place only to wither when he was back in New York. What could have been so nourishing to his spirit in the atmosphere and events of this year in Denmark?

A number of circumstances combined to produce the most benign and stabilizing influences that Noguchi was to encounter at any time during his life. He was under no pressure to meet deadlines in his work; he could proceed at his own pace with help and guidance offered but not imposed on him. The outbreak of the Russo-Japanese War provided him with unusual opportunities to release emotional steam, which had been held under great pressure for many years. He found friends, among them a young American, and he even fell in love. With more money than he had ever had, Noguchi almost kept his budget in balance. His only source of anxiety came from his inability to break his engagement without too much strain on his conscience and on the tolerance of his Japanese sponsors, Chiwaki and Kobayashi.

Noguchi entered the Scandinavian community of scientists who had a tradition of successful investigations that were stimulated by immediate practical problems, an approach that he liked. William Bulloch writes, "E. C. Hansen, working in the Carlsberg Brewery in Copenhagen, in a long series of publications dating from 1883 onwards . . . greatly improved the dilution technique of raising pure cultures of yeast and indeed revolutionized this aspect of the brewery industry."[5] Armauer Hansen of Norway had discovered the leprosy bacillus and was an authority on this disease which is now often referred to as Hansen's Disease. Bernhard Bang found the cause of abortion in cattle (*brucella abortus*), also known as undulant fever. Svente Arrhenius of Sweden, who was awarded the Nobel Prize just after Noguchi's arrival in Denmark, was a chemist who was actively interested in many fields, among which was the chemical nature of toxin antitoxin relationships. Carl Julius Salomonsen, the first director of a sero therapeutic department at the University Laboratory for Medical Biology in Copenhagen, which had been started in a small way in 1894, started a course in bacteriology at that same time. This course has been described as "the first in the world of its kind," and Thorvald Madsen was one of the original students in the class.[6]

Madsen had diphtheria as a child.[7] His older sister had died of the disease, and he is said to have been influenced by these personal experiences to study medicine. He made studies of diphtheria for his doctor's thesis. The disease was occurring with great frequency at that time with a high mortality rate. The development of a diptheria immune serum for treatment and a diptheria antitoxin for prevention by von Behring in Germany in 1890 had created a large demand for the immune serum and antitoxin. However, when Madsen received his doctor's degree in 1896, many problems of manufacture were still unsolved. Laboratories in France and Germany were engaged in a frantic competition to gain supremacy through the solution of these problems along with their efforts to understand the general nature of antitoxin reactions.

During the five years following the award of his medical degree, Madsen worked on these problems in Copenhagen and spent time in Ehrlich's laboratory in Berlin, at the Pasteur Institute in Paris, and at Arrhenius's laboratory in Uppsala. Madsen was twenty-six years old when he received his medical degree, and by the time he was thirty he probably knew as much about the problems concerning the diphtheria immune serum and antitoxin as anyone in Europe. He had even attended a congress in Russia during this period and had taken advantage of the opportunity to learn something of the diphtheria problems there.

In Copenhagen, the university was making some diphtheria immune serum, but even though the university's output was supplemented by imports from Germany, the amount was far from adequate to meet the demand. When the university administration came to the conclusion that large-scale manufacture of immune serum was not an appropriate function for a university laboratory, the medical profession gave its full support to the establishment of a serum institute. Professor Salomonsen was appointed director of the proposed *Seruminstitut* (a part-time position) and young Madsen was appointed to the full-time position of director of the laboratories. In spite of his youth, Madsen's personal attributes and knowledge of diphtheria were sufficiently well known and respected that his appointment was not questioned.

Madsen's appointment took effect in the spring of 1901. During the summer of that year, while the physical facilities were under construction, he made a tour of the laboratories of the United States and Canada. He contracted typhoid fever while he was in the Philadelphia area and was admitted to Johns Hopkins University Hospital where he and Dr. Welch became acquainted. Madsen had come to learn how the new hospitals and research facilities were being developed in America, and Dr. Welch, who was taking an active part in the design of the proposed Rockefeller Institute, was interested in discussing the latest developments in Europe. Since Madsen's illness was relatively mild, he and Welch used the period of his hospitalization for more extensive discussions than might otherwise have been possible. Madsen was a gentle man, attractive in his youthful appearance and enthusiasm, and well informed.[8]

The *Seruminstitut* opened in September 1902, less than a year before Noguchi's arrival in Denmark. For the first few months, Madsen left Noguchi free to finish reports of his previous year's work, and to start experiments of his own choosing. Noguchi was the first foreign visitor, and this circumstance, together with the prestige of his sponsors, Welch and Flexner, Carnegie and Rockefeller, guaranteed him a warm and respectful acceptance.

During the first few months, Noguchi wrote to no one but Flexner, and in these letters he spoke of little but his work. He expanded only to tell of a visit by Arrhenius and of the latter's Nobel Prize, of Madsen's being away or very busy, and of his finding the darkness of the Danish winter somewhat disagreeable. Arrhenius spent a month in Copenhagen during which time he once suggested that Noguchi's interpretation of one of his experiments was mistaken. Noguchi flared up in protest, writing Flexner that he was sure that he was right and that Arrhenius was wrong, but he soon dropped the matter.[9]

The accomplishments of Noguchi's year in Denmark, as of all his other work to date, were of no monumental importance. As a result of the year's work, he published five papers jointly with Madsen, and three or four of his own. The papers were products of painstaking, detailed accumulation of data on toxin-antitoxin reactions, and on the influences of various physical and chemical factors on different components of these reactions. Noguchi made some attempt to produce a snake venom antidote, but beyond the knowledge and technique he acquired, his work was still part of the routine required to lay a foundation for future research in immunology. He wrote Flexner that the fulfillment of his desire not to disgrace the new institute in New York would require more than his whole energy; whereas this may have been his polite expression of humility, it did replace the bragging daydreams.

During the first week of February 1904, the Russo-Japanese War broke out. Noguchi wrote Flexner:

> Since the outbreak of the war between Russia and my country I am spending every day with great anxiety and excitement. I cannot put the war news aside without reading every word of many papers — and am now reading the Danish, English, German, and sometimes French newspapers. It could have been settled better than this; I am very sorry. I fear and do not wish to see the result of this struggle.

After reporting on his work at some length, Noguchi returned to the subject of the war:

I am sorry that I cannot grasp clearly the meaning of your opinion about my return to Japan, or about my duty as a Japanese at this time . . . .I believe that it is my duty for my country, to strive to the end to accomplish my scientific studies, in which I am devoting myself at the present. . . . I am sure that I cannot be any better help to my country than I can be in the scientific struggles. It is entirely unnecessary to go back to Japan; I will remain as a scientific soldier for medicine.[10]

It is unfortunate that the letter from Flexner which elicited this response is not available. Flexner, the "protocol man," had probably raised the question of Noguchi's return to Japan only as a courtesy. Noguchi's reaction, however, was one of panic. He was horrified at the thought that Westerners might expect him to give up everything, just as his goals appeared to be within reach, to return to Japan and offer his services, which he had every reason to believe would be received only with contempt. He rationalized his feelings into an explanation for remaining in Denmark, and he now began a series of lengthy letters to his usual correspondents in Japan, Chiwaki and Kobayashi.

Noguchi's major concern was the war. In the first letter of this series he told Kobayashi, "We are approaching the most important moment in the history of our country." He worried that, since the reports in European newspapers differed widely, he could not be sure which report was correct. He described the reactions of the European press and his feelings toward them, and ended by saying, "I am still studying hard. It is a citizen's duty to achieve something and let the name of his country be known to the world. Since I could not help Japan even if I were to return now . . . I intend to stay here and do my best in my research.[11]

The second letter of this series, written to Chiwaki four days later, showed a little more assurance.

I feel very confident of the Japanese navy since hearing of frequent victories since February 8 in the Sea of Ryojun [off Port Arthur]. The whole of Europe is as if struck by lightning, and whenever a Japanese victory is reported people look doubtful, jealous and surprised. . . . Generally speaking, Europeans pray for a Russian victory because of their racial feelings. A great many newspapers seem to be awaiting the results of a great land battle. They predict a Russian victory because the Russian army is much larger than the Japanese. Although they seem to know that the Russian navy is no match for the Japanese, they do not realize how effective the Japanese army is. What will they say if the Japanese army wins the land battle also! They will surely say the savages of Japan are fit only for battle. They often look down upon us and say that we are lacking in moral standards. It may appear that way, but Japanese people never act in an immoral way. Japanese soldiers have never raped or murdered or stolen as the Europeans have, and Japanese have never plotted behind people's backs like white people do. They often say that morals between Japanese men and women are low, but

how about themselves! There are mistresses and lovers and adultery and
prostitution everywhere. They may think the Japanese do not know the
true situation, but it is easy to see what they are unless one is stupid.[12]

In a third letter, written almost two months later to Chiwaki, Noguchi's
excitement appears to have been sustained but the Danish reactions gave
him some pause.

> . . . in German and French commentaries there is frequent mention of
> the yellow peril. . . .Denmark is only as large as the palm of one's hand so
> they don't have any political power, but they comment on the war as all
> people do. Generally speaking they are pro-Japanese, but they don't
> make it very decisive because they are scared of the Russians. As you
> probably know, the wife of the Russian Emperor is the second Princess
> of Denmark, and the wife of the British King is the first Princess of Den-
> mark, which makes them all relatives. . . . The other day the Crown
> Prince of this country (brother of the Empresses of Russia and England)
> visited this laboratory, and the next day he invited all of us to the Im-
> perial Palace for dinner. They are very democratic and we all mingled
> and talked and stayed for about two hours. . . . Please send me
> magazines which have articles on the war.[13]

For Noguchi the war was at first a source of uneasiness and anxiety, but
with the success of the Japanese forces it became an invitation to him to express his righeous wrath, which could not have been more timely, and he
enjoyed every minute of it. For the first time in his life, he had been daring
to hope that his struggles might be at an end, that he could work at his own
level and have his work appreciated and accepted. In what he perceived as
the democratic atmosphere of Denmark, after his one flare-up in response
to the correction by Arrhenius, he talked no more of his superiority to
others or of his work being singled out for special attention. In the past,
during periods when Noguchi had met with frustration and despaired of
his future, he had spun wild dreams of becoming physician to the emperor
of Japan and had written scathingly of the Japanese social stratification
and cliques. Now, as the position and acceptance he craved appeared to be
within his reach, he was able to detach himself from the role of the
denigrated individual and identify with his countrymen, protesting their
denigration as a group: ". . . whenever a Japanese victory is reported, people look doubtful, jealous, and surprised."

Before he had left the United States to go to Denmark, Noguchi had
written Chiwaki that he planned to remain in New York, and to Kobayashi
that he now represented American rather than Japanese medicine. His life
and hopes were clearly identified with the West. But some of the Western
reactions to the war brought home to him what he already knew and
resented: that no matter what distinction he might achieve, among certain
people even in the democratic countries of the West he would always carry
some stigma as a foreigner, and particularly as an Oriental. Noguchi now

channeled the accumulated rage of his own past into broader resentment of this broader prejudice and exulted over the humiliation of the white man at the hands of Orientals even as he could maintain some detachment from the war and enjoy his life in the West. He had progressed to an alliance with his own people in the unfair treatment they received as a group rather than raging alone as an ambitious peasant at the bottom of the society of his own country.

There was another matter that Noguchi placed in a secondary position

Hideyo Noguchi, Copenhagen, 1904. University of Pennsylvania Archives

in these letters, but which had continued to worry him ever since he first mentioned it in his letters from Philadelphia. Although Noguchi had written Kobayashi that he had taken steps to break his engagement to the Saitō girl, he had thought about it during his vacation in Schwenksville and was reluctant to pursue this course. He had written twice of his sympathy for the girl and his obligation to her, to her family, and to Chiwaki. Even when "courage was boiling" in his mind as he sailed from America, he could not decide and had promised to consult Kobayashi again later. But he did not write Kobayashi again for more than three months. He could not skip a letter at New Year's time, but dealt with the engagement only in a postscript. Still undecided, Noguchi noted that his mother had urged him to break the engagement; but he was equally undecided about the suggestion of Kobayashi's niece.[14] He discussed the matter of the engagement in a long letter to Chiwaki; as with his old attraction to Yoneko, he felt strongly drawn to the girl who had suffered in a manner similar to his own. Beyond all other considerations, it is likely that Chiwaki had reminded him of his obligations.[15]

Noguchi suddenly stopped writing about the engagement, and we detect hints that another love interest had taken the place of his fiancée in his mind. Okumura writes that when he visited Noguchi in New York at Christmastime in 1904, after Noguchi's return from Copenhagen, there was a picture of an attractive Danish girl hanging on the wall of Noguchi's apartment, and that Noguchi was hoping to marry this girl. Other Japanese students who were in New York at this time knew of Noguchi's affection for this girl who was variously reported back in Japan to have been the daughter of an army surgeon and the daughter of Noguchi's landlord in Copenhagen.

Dr. Madsen's father was minister of war in the Danish Cabinet at the time of Noguchi's arrival in Copenhagen, and Noguchi had written Chiwaki that this circumstance had made it possible for him to live in an artillery camp. There was an artillery camp adjacent to the *Seruminstitut*, and an apartment building at the edge of the camp ground in which non-commissioned personnel could not only rent apartments but also take in boarders. Noguchi lived in one of these apartments with the family of a sergeant; the title "sergeant" became "surgeon" as the tale was repeated. No special arrangements, much less arrangements made by a Cabinet Minister, had been required for Noguchi to live in this place.[16] Could Noguchi have made this mistake? It seems more likely that he wove a tale out of fantasies that grew as he identified with the military glories of his country at the same time as he fell in love with the daughter of his landlord, the sergeant.

We do not know exactly what did take place. At some ill-defined time, the father of the girl is said to have given Noguchi a sword to symbolize the severing of the relationship. The picture of the girl was gone from Noguchi's apartment wall when Okumura visited again the following summer and has never been seen again; but a Danish sword was among Noguchi's possessions when he died and it is now on display in the Noguchi Memorial Museum in Inawashiro. Several of his Japanese friends in New York told, on their return to Japan, of Noguchi's asking people which they considered more valuable: a sword that was given him as an honor (referring to a sword which he later received with a commission in the army of Ecuador in recognition of his work there with yellow fever) or a sword that was given him as a symbol of love (referring to the Danish sword).[17] We know that Noguchi's love for the Danish girl came to naught, but it undoubtedly contributed to the happiness of his year in Denmark.

Toward the end of March 1904, a medical student from the University of Michigan, L. W. Famulener, arrived at the *Seruminstitut* and was assigned to share the laboratory in which Noguchi worked. They got along well. Noguchi wrote Flexner that he was happy to be working alongside an American again, and he wrote several friendly letters to Famulener during the following years. Among the letters that Noguchi wrote to many people during his life, the letters to Famulener can be compared, in their straightforward, informal style, only to those he wrote to the mysterious Kodama in whom Noguchi was so willing and able to confide. One of the first things Noguchi did on his return to New York was to get a five-pound bank note and send it to Famulener in repayment of a loan.[18] Noguchi was not yet in control of his finances, and the sum he had borrowed was considerably larger than five pounds. For Noguchi to repay any of this loan so promptly, however, was definite and conspicuous progress. It is also the first record of Noguchi's ever having borrowed from a Westerner, and it may be that pressure to repay was applied on him by Flexner.

For the Copenhagen period, it is significant that Noguchi had been able to form the kind of friendship which the later letters demonstrate. Although the friendship was no doubt encouraged by their being the only two foreign guests at the institute and by Famulener's friendly personality, it must also have been a reflection of some increased measure of security in Noguchi. There is a real sense of the "classmate" feeling as he wrote Famulener later "I am inclined to think that you had studied under the best teacher and at the best Institute in Europe."[19]

In Hamburg, en route back to New York in September, Noguchi picked up his mail from Japan at the offices of Mitsui, and learned of the sudden death of Chiwaki's little daughter of bacillary dysentery. He wrote condolences to Chiwaki in the most flowery language but, on a practical level, he considered the real danger to the rest of the family and their need to have something specific to do in their time of sorrow. He told Chiwaki that the disease organism had been discovered by Shiga, who was at that time in

Germany but would soon be returning to Japan. He discussed the con-
tagion and measures that could be taken to protect the other members of
the household, and suggested that Chiwaki talk with Shiga when the latter
returned in October. Noguchi mailed this letter just before sailing for New
York, noting that it would go east, via Siberia, as he went west.[20]

Whether Noguchi ever recognized the extent to which Madsen had supported him while he found his way, and had diverted him from grandiose
plans for the manufacture of antivenom without discouraging or
disparaging him, Noguchi had found a comfortable and appropriate pace
that took advantage of his ability to work long and patiently, with
meticulous concern for detail if not order. Noguchi seemed to gain as
much satisfaction from it as he ever did when he was later involved in more
ambitious projects. It is probably the real tragedy of his life that he was
drawn into the competitive struggle to free the world from disease; if his
objectives had remained less pretentious, his contribution would likely
have been greater in the long run. He said that he learned accuracy of
technique from Madsen, and perhaps he did. He had been impressed by
the cleanliness and order of the Danish laboratories, but he never became
an orderly worker. For this one year, however, he surely enjoyed and prof-
ited from Madsen's guidance.

If these are the implications for Noguchi's work, what can one say of his
personal life and the combination of events that made it in total such an
outstandingly happy year in his life? Even though he could never admit to his
correspondents in Japan that he was enjoying life in the West, even though
he made full use of the Russo-Japanese war to castigate Westerners for
their anti-Japanese attitudes, he made it clear that he planned to remain in
the West and that he found some Western people sympathetic. And a
modicum of the accumulated bitterness, anxiety, and fear of deprivation
seems to have been dissipated as the door to personal friendship, and even
to a relationship of love, had begun to open.

10

Anxious Beginnings in New York

THE DAY NOGUCHI ARRIVED IN NEW YORK, SEPTEMBER 22, 1904, HE WROTE to Famulener in Copenhagen:

> I arrived here safely this afternoon and took my first American meal at this hotel. Well, it tasted something like American, and I thought of you at once. Poor Famulener, I thought, is even robbed of his daily necessities, and yet for his own future.[1]

Shortly thereafter, as noted previously, he sent a five-pound bank note to Famulener explaining that preparations for getting the new institute started had kept him from sending it immediately upon his arrival.[2] He wrote still a third letter to Famulener on October 26 in which he referred to his having sent the money. His repayment of this loan seems to have impressed Noguchi. He recalled places and good times in Copenhagen as though he hated to part with the life he had known there.[3] By the time he wrote to Chiwaki in November, he was uneasy:

> The month between October 1 and November 1 was spent in preparation of equipment, books, materials and animals for my research. . . . There are six members of the new institute including Dr. Flexner who is the Director. Two of them were already professors and they are over forty years of age. Their positions are above mine, but their departments are separate from mine. The other three are in positions below mine although age-wise, I am the youngest. . . . Living expenses are double here from what they were in Philadelphia. My annual income is $1800. . . . Dr. Flexner told me to save some money for future emergencies and I intend to do so, but after the travel, etc. I don't think

I can save any money this year. I am determined to save, if only a small
amount, starting next year.

I have already given up drinking and smoking and I do not feel like
having any. (Of course I stopped drinking altogether when I came to the
U.S.) I am still doing research on snake venom.[4]

By January his anxiety was in full bloom. On the first day of that month
he wrote Kobayashi that he had not had enough money in Europe and had
borrowed 600-700 yen; he had paid back all but $50 since his return
(which would mean $250-300 out of his three-month total income of $450,
if he really had done this.) Okumura had arrived in New York in
September to study dentistry, and Chiwaki had offered to pay the debt to
the Saitō family if Noguchi would contribute $35 a month to Okumura's
support. Noguchi had agreed to do this, but now he told Kobayashi that he
could not continue. "I have already decided that I will stop assisting
Okumura from March on, so I cannot ask Chiwaki to pay the Saitōs."[5]

Noguchi's letter to Chiwaki a week later mentions neither Okumura nor
the Saitōs directly: "I checked my financial status after coming back to
the U.S. and I don't have any money left at the end of the month. I realize
that I must be independent and save for the future to provide for emergen-
cies." He said that he was suffering such severe pain from hemorrhoids that
an operation was required, and

> this is the reason that I have to save for emergencies. Since I am being
> treated like a gentleman, I must act like a gentleman; and I must be
> especially respectable since I am always with Americans. I cannot ask
> others for money — Dr. Flexner advised me of this a while ago. The fact
> is that I am living among complete strangers and not having money sav-
> ed for a day's illness or medicine puts me in the position of a lowly
> laborer who is forced to live that way. I have come to realize that this is
> not the way for an independent man to live. . . . When I was being given
> money for clothes, I had no problems in getting new clothes every
> season; but now that I have been on my own since last September I must
> take care of this myself. Especially because of the expenses of my ocean
> voyage and my travel in Europe, I have many debts which I have not
> finished paying yet. I am ashamed that I haven't bought any winter
> clothes since last year. I have no coat and am using a raincoat. . . . I
> haven't been able to find places for installment buying so it's awfully in-
> convenient for a person like me.
>
> As you know, it has been four full years since I came to the U.S. and
> during the first three years, although it was called a salary, I received
> only an allowance for maintenance. . . . Since $25 was so disheartening
> to say, I didn't tell you the truth. That just paid living expenses. As I
> have told you, this institute is very rich. This year (since last September)
> I have been receiving the salary of an average person. Since I don't have
> much experience, I have been trying hard to manage on that income
> but I don't have any extra money. This unpleasantness has reached its
> greatest extreme since my arrival in the U.S. I would like to be more
> frugal but my dependent life of the past has almost become my nature

. . . .I didn't sleep very well last night and my brain is unclear—everything is bothering me.[6]

In Feburary, 1905, when he wrote Famulener in detail about his hemorrhoids and an operation that had greatly relieved him, Noguchi seemed in better spirits:

> You speak in your last letter about your letter to Prof. Ehrlich. I cannot imagine why he does not answer your letter, but it is not unlikely that there is something in his attitude against the Arrhenius-Madsen school, do you not think so? I really cannot see real necessity of your entering the Ehrlich's after you had studied under Dr. Madsen. I believe you have already learned every fundamental technique in our "serumline" and would not cut too much ice in Frankfurt.
>
> Theories are not to be taught by anybody outside of ourselves. We are the best teachers of the truth—I mean by this that we ought to convince ourselves chiefly by our own experiences and own experiments. Nobody can teach you beyond the methods of investigations, and you have gotten them over there. I am inclined to think that you have studied under the best teacher and at the best Institute in Europe.[7]

In April Noguchi told Kobayashi that he had finally asked Chiwaki to break the engagement and had heard that the "Saitōs have accepted." Chiwaki had returned the dowry and Noguchi conveyed his mixed sense of relief and regret: "I will return the money to Chiwaki later. It has been five years that this arrangement has dragged on but it is finally broken. . . . If I end up being single all my life, it will be the will of heaven."[8] Three days later he confessed to Chiwaki that he was miserable:

> Although I am still in the confusion of the crowded city, I have returned to my earlier self and all kinds of emotions have been revived within me. Since I have settled down in New York I haven't had one day of leisure, or even if I have one I can't be calm and think about fundamental things because I am an ordinary person in the midst of frantic endeavor so it's unbearably unpleasant. The inhuman state here is such that I don't even recognize my landlord's face even though I have been here half a year. There is always worry of robbery if one is not careful; human life is no better than an animal's; life in large cities is degraded. In addition to all this, I am suffering from my hemorrhoids which depresses me terribly.*
>
> I was asked last month to write a chapter on snake venoms for a textbook in internal medicine. The deadline is already set so I have to get this done, and in the meantime I have to work at the Institute during the day and I am pushing my brain for a new experiment so I don't have any leisure.

He was worried about the political settlement of the Russo-Japanese War, and he was worried about his family. He said he was working on

---

*Even though he had told Famulener two months before that he had recovered.

trachoma but warned Chiwaki that this must be kept secret, and he continued:

> When I think of it, I am no longer at an age when I can be dependent
> on others. I feel guilty for being what I am when a person of strong will
> could already be taking care of a student or two. Dear teacher, please
> overlook my position for a few years more. Someday soon I will raise my
> head.[9]

Following this Noguchi did not write a single letter to anyone for eight
months, until December when he wrote to Famulener who had now returned
to Michigan:

> Indeed I am feeling myself almost like a criminal in not having
> answered your letters both from Chicago and from Michigan. I am
> ashamed of myself of my long neglect of correspondence. I have not
> written a single letter to any person since last June. I was, in fact, unable
> to turn my thoughts to anything but work. It was a kind of mental fit
> and I could not resist it. I am just now somewhat free from my pressing
> burden and write this note to you. I wish you will kindly excuse me of
> my impolite manner (it has all unconsciously happened!).[10]

Noguchi told of trying to work with trachoma, but made only casual
mention of a project that proved to be of major importance in his later
career: "I made some studies on the occurrence of spirochaeta pallida in
syphilis with positive result—together with Dr. Flexner, but I am not working on this any more." He had written a chapter on venoms for a textbook[11] but said, "It was a very hard task for me on account of the
language and took all my time and brain away from me for a time!" He
told of translating into English two papers that he had written in French
while he was in Copenhagen and continued his complaints:

> This is a hard job for me, and bothered me so greatly that I have neither
> energy, nor ambition after spending hours after hours on such things
> like that. I can do work, but cannot write the results well. . . .
> Manuscript writing is the worst task for me. Please imagine that you
> write a paper in German or French after a stay of only a few years. This
> is the case with me with English."*

Listing still other projects that he was trying to establish in the
laboratory and that took a lot of detailed work, Noguchi burst out with his
total feeling of frustration:

> I am almost exhausted and I feel the weight of my situation, because
> every one working at this Institute is expected by the outsiders to do
> something. Yet, as you know, we cannot find a new thing every day!
>   It will be a great thing if one can do his own way without being com-
> pelled from the top. A research work can be done only by a thorough

---

*Flexner stated in his Biographical Sketch after Noguchi's death that Noguchi always
wrote easily and quickly.

and free thinking. . . . Now you see why I could not write a single letter since last June. Neither have I written to my people in Japan, nor to my friends or teachers abroad! I feel myself pity! I don't like this city at all![13]

He was caught. He had lost all hope of being married either to his Japanese fiancée or to the Danish girl; he hated the lonely, impersonal, big city; he realized what it meant to have to manage his own affairs, financial and otherwise, and how ill-equipped he was to do so. He joined the competitive scramble to publish as much as possible and as rapidly as possible, and with it he again became secretive and anxious, losing much of his assurance. He struggled with the truly enormous task of writing papers, even postponing the writing of the monograph on his snake venom work for which he had a commitment to the Carnegie Institution.

Noguchi's reports of his work to Famulener are a fair statement of his accomplishments and failures. Between the years 1905 and 1908, his list of twenty-eight publications included the last reports of his work with snake venoms and the routine observations of immunologic relationships, studies on tetanus, and the chapter on venoms in Osler's and McCrae's *Modern Medicine*. Most important of all was the study of the occurrence of *Spirochaeta pallida* in syphilis, which he had mentioned to Famulener only in passing.[14]

Schaudinn and Hoffmann, in Berlin, had announced their discovery of *Spirochaeta pallida* in 1905 and its causative relationship to syphilis. Flexner set Noguchi to the task of confirming their findings as soon as the announcement appeared, and Noguchi did so with speed, accuracy, and aplomb, but he returned to his other work.

In 1906, however, when the Wassermann reaction test was announced, Noguchi recognized its potential both as a diagnostic tool and as an opening into the kind of investigation he felt best prepared to undertake. This test made use of the serologic reactions with which he had become so familiar in his snake venom studies. The development of a reliable serologic test for syphilis was an objective that could utilize his particular skills and experience, one which could make major contributions to a "big" problem, and one which would have immediate practical application.

Noguchi's first paper on the subject appeared in 1909, but since he published twelve papers in this field during that year (in addition to the snake venom monograph) it seems fair to assume that he started working in this field shortly after the reports of the Wassermann test appeared in 1906.

Meanwhile his personal life became increasingly disorganized. He had found a boardinghouse when he first returned from Europe and was still living there when Okumura paid a last visit before returning to Japan in the summer of 1906. Okumura tells of one of the other boarders asking at breakfast: "Is there a Japanese in New York by the name of Noguchi?" adding that the name was in the newspaper. Since Noguchi had lost his calling cards somewhere, and since he had been drinking at the time, he became frightened as he dreaded to hear where the cards might have been found. (The newspaper story was of another Noguchi.) From his own observations, Okumura knew that Noguchi drank and smoked a great deal at this time, and he marveled at the extent to which Noguchi could get along without sleep.[15]

In September 1906, Noguchi moved into a flat on East Sixty-Fifth Street with Tatsutaro Miyabara, a young man who had come from Japan to study medicine in New York. They had decided to set up a medical practice for the Japanese community of New York from this apartment, in which there was a laundry tub but nothing else. They even had to put quarters into a gas meter to fuel the stove that they bought in a twenty-month installment purchase. They announced the opening of their clinic in the Japanese-language newspaper, but received little patronage, since a Japanese physician was already well established in the community. Their hasty decision to start their clinic bothered Miyabara, but not Noguchi.

Miyabara found other episodes to be indicative of Noguchi's attitudes during this period. On a walk in the park, Noguchi stopped before the lion's cage in the zoo and said: "Watch the actions of a predator. It will help you some day." Another time he told Miyabara, "I will work like an obedient lamb and will be devoured after all as food for the lion."[16] As Miyabara prepared to return to Japan in August 1907, Noguchi was envious. He kept asking about Miyabara's plans, putting needling questions to him. "How much money do you think you can make?" When Miyabara answered 300,000 yen, Noguchi exclaimed, "Is that all! Say a million yen even though it's a lie! But I don't need such a petty sum. I would rather stay in the United States and show the American people, who have been against the Japanese lately, just how much a Japanese can do." When Miyabara remarked that this sounded like an extravagant statement, Noguchi defended himself. "Research is a rather speculative business. One may hit it big or not, even though one works all his life. . . . If I can't achieve anything after working so hard, I may end as a railway suicide."[17]

Shortly after Miyabara left, a young dental student, Norio Araki, was referred to Noguchi by Chiwaki, and Noguchi had him move in at once. Araki found the apartment practically empty, and he busied himself with getting some cooking utensils and various furnishings to make the place a little more comfortable. This younger man was obligated to even a slightly older man who took him into his home in this way, and Araki took his obligations seriously. He got up and cooked breakfast so that Noguchi could be at the institute at the time he was supposed to arrive—eight o'clock—and tried to have dinner ready when Noguchi returned in the evening. Of the first meal that Araki cooked, Noguchi said it was too salty;

of the second, that the rice was burned; of the third, that something was wrong with the potatoes. Each time they threw the food away and went out to eat. But this period passed. Noguchi read during meals, and Araki was surprised that he could notice anything wrong with the food since he seemed to pay so little attention to it.

Araki came home late in the evening after his classes and went to bed around midnight, but Noguchi was never in bed at that time. When Araki got up at 5:30 in the morning to start cooking breakfast, Noguchi was always up and at work. Araki found Noguchi's attitudes expressed in the selections that Araki quoted: "if one is doing something, someone else is bound to be working on the same thing. If I can make progress even one second faster, I must do so." And, "You [Araki] are young. You must make up your mind before the age of forty, since your future will all be written by that time."[18]

Working with his microscope at night, Noguchi smoked large cigars that he considered a stimulant to keep him awake. He explained to Araki why he had to do everything himself when Araki offered to help with grinding tissue for slide preparation. He said that Ehrlich worked with his head and thus could have others do manual things and make drawings for him; but Noguchi trusted his own hand better than he trusted his head. Noguchi even washed his own test tubes. He lectured Araki, saying "If you want to be liked by your boss, never say 'No.' " He resented putting Flexner's name on papers when he felt he had done all the work himself. "If I had said 'No,' I would have been fired a long time ago and couldn't have continued my work."[19] So even though this was annoying, he did it. He urged Araki to publish, and even offered to write papers for him in English as soon as Araki brought home his results.

Noguchi needed human blood for his spirochete studies, and he not only used his own but began to ask for contributions from Araki. The first ounce was all right, but as the demand continued, Araki lost his enthusiasm. Noguchi did not want to be interrupted while he was at work so Araki often communicated with him by leaving notes. Soon Araki began to find notes from Noguchi saying, "I need five dollars" or "I need ten dollars." and Araki would leave the money. But then it became ten or twenty dollars; the total reached a few hundred dollars and Araki decided to call a halt. Although he was not able to make his protest as politely as he had wished, Araki nevertheless said that he felt that it was not good for Noguchi to go on borrowing money in this way and he wanted the whole sum repaid. Noguchi said he would do so, but it would have to be in small installments and Araki would have to come to the institute every pay day and Noguchi would give him a payment. This was the only way he could do it. Araki came to the institute. He had really hoped to improve Noguchi's way of living, but settled for just getting his money back.

Araki persuaded Noguchi to move to a smaller, more practical apartment than the one that Noguchi and Miyabara had rented for their medical practice. Since the new apartment was less expensive and close to the institute, Araki's arranging the move was a real service to Noguchi as were the many other things he did to make Noguchi's life easier. But there is no sign that he got any thanks for any of it. If Araki was aware of Noguchi's drinking or any other form of dissipation, he did not say so.

Noguchi had known even before he left for Denmark that he was to be awarded an honorary master's degree by the University of Pennsylvania. On July 19, 1907, he wrote Chiwaki:

I have received an honorary degree of master of science from the University of Pennsylvania. . . .Everything is beautiful when it is still in a dream state, but when it becomes a reality it is no longer interesting to me. . . . "When one wish comes true, another is born." . . . Now I intend to request a medical degree from the Japanese government. . . . I don't know what my name means to Japanese scholars, but in the medical society of Europe and America I am the most widely known Japanese with the exception of Kitasato and Shiga.[20]

For the next few years, Noguchi sent reprints of all his publications to Chiwaki, asking him to submit them to the Ministry of Education as research reports in support of his application for a medical degree. As granted in Japan, this degree was the equivalent of the German *privat docent* in that it was based on achievement in research and carried academic credentials, which an American M.D. did not. Araki was impressed by the large number of publications that Noguchi submitted in support of his application, but Noguchi insisted that he did not care about the degree—he was just trying to get it to make his mother happy. His real attitude may be detected in a letter he wrote to Kobayashi in 1908:

My PhD dissertation was submitted the other day and after a few years I will be permitted to have the degree. Whatever the University or the Japanese government says, my work, which is recognized throughout the world, deserves more than a doctor's degree. Even if one has a degree and a few titles from Japan, it doesn't mean a thing if nobody pays attention when one comes abroad. . . . It is very funny that some people, who never learn to speak German during their few years study in Germany, should receive a doctor's degree as soon as they get home. They are not at all known in the medical society of the world. It would be an insult on the part of Japanese medicine toward the world if these men should spend their pretentious lives holding up their time of study abroad as a golden trophy. As for me, fortunately I don't need a doctor's degree and even if I had one, it wouldn't mean anything in European or American eyes. Nobody respects Japanese degrees so I don't give a damn about getting one.[21]

At the same time as he was expressing such contempt, Noguchi was presenting a confusingly large number of papers on a variety of subjects.

Ultimately, four papers were selected and collectively designated as his dissertation.

From the time of his going to Denmark, through the first few months after his return to the United States, Noguchi had been optimistically looking forward to making his career in the West, only visiting Japan when he chose to do so. Under the pressure he found in New York, he started thinking again of forcing his way into one of the Japanese academic cliques. He hated to beg for a degree, but he knew he could not expect an offer of an academic position unless he held a degree from one of the prestige universities of Japan.

Noguchi knew that if he were successful in his efforts to become the world expert in serum diagnosis of syphilis, and perhaps expand into the treatment and prevention of this disease, his name and services would be valuable anywhere. As he labored to achieve this objective in his work, he carried on his ancillary program of getting degrees and adding to his list of publications. When the medical degree was finally awarded by Kyoto Imperial University in February 1911, Noguchi immediately made application for a Ph.D. in science from Tokyo University. Even though it was equivalent in status to his medical degree, he was leaving no stone unturned. But by the time the second degree was granted, he and Araki were no longer living together. He told Araki he was moving out because he was going to be married.

11

His Family and His Marriage

BOTH OSHIKA AND KOBAYASHI WERE WILLING ENOUGH TO BEAR WITH Noguchi while they knew he was living on student stipends, and even beyond that time, as long as they still had some reason to hope that he would come back to Japan and settle down with a Japanese wife of their choosing. But as it became clear that Noguchi was getting a good salary at the Rockefeller Institute (an enormous income by their standards) they ran out of tolerance for his doing nothing to help with the inevitably recurring problems of his family. When rising costs and crop failures added still further to the Noguchi family burdens, Kobayashi decided that things had gone far enough and it was time for him to take a more aggressive stand with Noguchi.

About the same time that Noguchi had come to the United States, Oshika had begun to help the local midwife in her work. She rapidly acquired such skill and popularity that when the government set up its first training program for midwives, Oshika was chosen as the candidate for training from her village. Most of the candidates were no more literate than she, so it was not too surprising that she passed the examinations and was licensed to practice. The villagers collected funds to buy her the required equipment—a watch and a stethoscope—articles no midwives had ever used before, and she now had a new dignity as a person with officially recognized status.[1] Although her earnings were probably never more than ¥100 ($50) a year, she received gifts of food and other practical help from families who could pay no fee. In her circumstances, even the most meager earnings would have helped. By 1903, she was even writing an occasional

letter to her son in America; in April of that year Noguchi wrote to
Kobayashi: "I heard from Nihei of Sanjōgata today and with this letter I
received a letter from my mother. It was all written in *kana** but I could
read it very well. I cried as I thought of my parents' life. . . .I am really
amazed that she wrote [this letter] herself. I didn't know she could write,
but I can't guess who might have written it for her."[2]

Although Kobayashi wrote that the year 1906 had brought a poor
harvest with considerable distress for the whole Tōhoku area, he added
that "poor as the Noguchi household is or may become, they don't have to
worry as long as I am alive."[3] So Noguchi continued to write about his progress and his efforts to live within his means without any sense of an
emergency at home. On July 9, 1907, when he wrote that he had received
the honorary degree from the University of Pennsylvania, he continued
"My salary may increase this October. . . .I have been getting $2000. . . .I
haven't saved yet, but I will soon save money and will return to Japan next
May."[4] At the end of July, however, Noguchi wrote Chiwaki that he had
been told that the family's situation was growing worse. "Therefore, I am
sending a money order addressed to you and would ask you to change it into Japanese money and send it to my mother. But please don't do it directly
but through Mr. Kobayashi. This is just my caution to prevent unnecessary
unpleasant feelings which might be aroused when an exaggerated rumor
goes around. As Mr. Kobayashi is a very careful man, he will handle this
very tactfully. . . ."[5]

Noguchi promised to send about ten dollars every month to his family,
and a month later he wrote Chiwaki that he had also bought some
obstetrical instruments and had shipped them to Chiwaki, again to be
transmitted to his mother.[6] It was a year before the instruments finally arrived, but when they came, Kobayashi wrote that Oshika was delighted to
have them. Even though regulations prohibited her from using them,
"they will strengthen your mother's self-confidence; your dutifulness will
impress people which will surely increase their trust in midwives."[7] Oshika
was never able to use the instruments, and she finally sold them to a physician who was pleased to have them.

Noguchi soon gave up the practice of sending his monthly remittances
through Chiwaki, but did maintain a policy of sending them to Kobayashi.
They all believed that it should not become common knowledge in the
village that Oshika was receiving money from her son. Everyone in the
village had been affected by the weather problems that were the basis for
the Noguchi family's troubles. The family had not been able to pay their
tax assessment for two years, and in 1907 they were notified that their land
was to be sold at auction.

---

*Simplified phonetic symbols for the Japanese syllabary that avoid the use of the more dif-
ficult Chinese characters used by educated people.

Oshika, in a frenzy at this intolerable prospect, managed to borrow enough money to postpone the calamity and Kobayashi issued his call for help from Noguchi. In meeting this crisis, they set in motion a series of negotiations that were to continue for nine years. Uniquely Japanese in their style, the negotiations were complicated (at least for a Westerner) by practices that had been held over from the customs of feudal days. The manner in which land titles were held, and in which tenant farming obligations were met and other debts were incurred and paid, all rooted in traditions of a feudal farming village, dominated the negotiations that Okumura reports in detail that a Westerner can neither follow nor comprehend.

Kobayashi was deeply involved. He negotiated, reported progress to Noguchi in long letters, and needled him when he was not helping enough. He not only loaned money to Oshika himself, he also bought property on behalf of the Noguchi family and held it until Noguchi should come home to take title to it, thus preventing the despised Sayosuke or Inu's drunken husband from having any control or being able to borrow on it. In 1908, Kobayashi wrote:

> As I have mentioned in every letter I have written, there is a world of difference between your hardship and your mother's hardships. Think of farming families in the past: most of them have never even seen fish except for a little trout at the year's end and some herring during planting season. One cigarette for an American gentleman would buy food for ten days for a Japanese farmer.[8]

Even Sayosuke had tried to step in and make his contribution during a period of special desperation. He had humbled himself to the extent of asking Mayor Nihei for a job, but was put down by the requirement that he deposit a fund for his drinking money.[9] He accepted a job doing odd chores around Kobayashi's house with a limitation on his drinking to two cups of *sake* every evening, but soon could no longer stand this restriction. He joined a migration to Hokkaidō, an area that had been largely unsettled because of its rigorous climate but was now being developed under government subsidy to so-called pioneer farmers.

In the aftermath of the Russo-Japanese War, and in the period of development that began in the northern areas of Japan at that time, prices had risen rapidly. Although this was the beginning of a period of relative prosperity, its benefits reached poor farmers slowly and they were not yet prepared to sustain themselves through the crises created by unpredictable weather. As Kobayashi described it, "Civilization has spread, but people are still faced with difficulties. It may not be a big mistake to say that civilization is fine for the rich, but to people below the middle class, of whom there are many, it brings suffering."[10]

For Oshika, therefore, any money she received from her son gave her an advantage in such times far beyond even its usual value. She and Kobayashi continued to conceal, not only from the villagers but from the family as well, the fact that Noguchi was sending—fairly regularly—the monthly payments he had promised and larger sums on occasion. Noguchi was told, in one letter, that Oshika was taking the money home only in small sums so that the family "would not become soft."[11] In the one letter from Oshika to her son, which is available and which is frequently reproduced and much beloved for its childish writing and ignorant, country-woman's language, she said: "I am not telling anybody that you sent me money. If I tell, all the money will be spent in drinking. Please come home soon, please come home soon, please come home soon, please come home soon."[12]

Another event in Noguchi's life in the West was to bring help to Oshika during this critical period as well as later. Noguchi had met Hajime Hoshi in Philadelphia and they saw each other occasionally during Noguchi's early days in New York. Hoshi, who was as highly organized as Noguchi was disorganized, had come from Fukushima Prefecture to study at Columbia University. After graduation, he started the first Japanese-American newspaper in New York, and sold it as soon as it was well established. Hoshi returned to Tokyo, where he started a pharmaceutical business from which he made a lot of money.

In a letter to Kobayashi in December 1913, Noguchi enclosed a letter he had received from Hoshi in which Hoshi had offered Noguchi some financial compensation for the privilege of using Noguchi's name as an "advisor" for his company. Noguchi suggested that Hoshi send the money to Kobayashi to be used for buying land or paying off debts for Oshika, and this was done.[13] Hoshi began sending Kobayashi fifteen yen a month in April 1915, and continued these payments until the time of Oshika's death in November 1918. This help came, of course, well after the time of greatest pressure both on the family at home and on Noguchi in New York, but it was an important part of the final achievement.

In spite of their very different modes of behavior, during all these years Kobayashi and Noguchi maintained a tiny but unusual observance indicating a basic agreement in their objectives for the Noguchi family. In the pledge of kinship they made when Noguchi was leaving Japan, Noguchi was granted permission to call Kobayashi "father." In 1903, Noguchi began to address the Kobayashis in his letters in the honorific terms used for parents.[14] With no words of explanation, in December 1906, Kobayashi began to refer to the Noguchi family as Jōka 城 家 ; in writing Jō 城 , he used the Chinese character meaning "castle," and for ka 家 he used the character for "household" or "family."[15] He thus created a word that is readily understood by anyone familiar with

Chinese characters even though the word does not appear in common use or in dictionaries. Noguchi picked this up in a letter he wrote in 1908 and referred to his family in this manner from this time on.[16] In 1909, Noguchi began to refer to his mother as *Jōbo,* 城 母 castle mother,[17] a custom that Kobayashi adopted in his reply and used thereafter.[18] Prior to this, Noguchi had spoken of his mother in the conventional deprecating form — *Gubo* — my foolish (or stupid) mother.

We dare not speculate too far in searching for subtle meaning within this wordless dialogue. It seems safe to note that it implied some basis for associating the Noguchi family with a castle, and supported the implication by using the scholarly (thus upper-class) Chinese characters. Beyond that, it appears significant that Noguchi and Kobayashi found themselves in such an easy collaboration at the same time as their letters dealt with day-to-day concerns. The restoration of the Noguchi family was eventually made manifest in Inawashiro by the outcome of Kobayashi's nine years of negotiation. In July 1916, he could rejoice that Jōka owned, free of debt, a paddy field and a second field (total about 5.38 acres), a house, a stable, and about three acres of mountain property; they had become, in his words, "a fine agricultural household."[19]

But before Kobayashi wrote these words, in the same six or seven years during which Noguchi was having his "mental fits" in trying to establish his position at the Rockefeller Institute while pressure from Japan was reaching out to bear on him, Noguchi was feeling his greatest need of a home and a wife.

As soon as Noguchi's engagement to the Saitō girl had been "honorably" disposed of, he asked Chiwaki to help him find another fiancée. Chiwaki sent him pictures and information about the family of a girl in Tokyo, but Noguchi unwittingly sent them to Inawashiro for approval at a time when conditions there were bad and Kobayashi responded only with sarcasm.[20] Noguchi was growing steadily more depressed in New York but Kobayashi responded to his tales of frantic efforts with tales of equally frantic efforts at home. Lapsing into a real holiday gloom, Noguchi wrote on December 17, 1908:

> Nothing has happened so far about my marriage, and I haven't asked about it either. For one thing, I don't have a minute's leisure from my studies, and for another, I am several thousands of Ri away from Japan. Even if I asked people in Japan to arrange a marriage for me, no one would risk arranging a marriage for a person he had never seen. I am at a loss. I am now thirty-one years old, have status and honor in this country, but I believe there is no use asking a matchmaker to find a wife for me. . . .I have been asking Dr. Chiwaki to look for my wife-to-be, but have received no answer. He probably has no time for others, being busy with mundane affairs. It is true — to conduct one's own life is hard enough. A man's strength does have its limitations and it is only natural that he

can't look after others as well. Going home at six o'clock after working all day with all my might in this strange land, exhausted to the point of being dizzy, finding my room dark with no light, and having to straighten my own bed, makes me feel very forlorn. A man's life is pretty gloomy. Being a foreigner, I haven't found any close friend which compounds my misery further. On the surface I am a fine scholar, but inside I am exactly like an exile. When I compare my life with my fellow American workers, everyone has a wife and is living a pleasant family life. Please let me know if you find someone who might marry me.

P.S. I found this letter dated December 17 and am sending it to you immediately.

April 3, 1901[21]

By the end of 1909 things had begun to improve in Inawashiro, and Kobayashi was encouraging Noguchi's efforts to get a degree from a Japanese university in the interests of a future career in Japan, and he even showed a grudging interest in Noguchi's ideas about marriage. "We don't mind a foreigner if she is a suitable person." But new floods brought new trouble, and in January 1911 Kobayashi wrote:

> As I have written before, the Kantō and Tōhoku areas are flooded. This is the greatest flood in history—it has been flooded for twenty days now and some paddy fields are completely destroyed. Jōka also suffered some damage. . . .They have been helped by the money you sent, but this is an emergency and I am asking you to cooperate. Of course, we are not starving, but we are in trouble.
>
> We sympathize with your ten years of hardship since going to the U.S. You must have contributed a great deal. We also sympathize with your sitting all by yourself under a lonely lamp three thousand Ri away from home. Please be sure to answer the following: 1. Please let us know, without hesitation, your feelings about taking a wife. 2. When are you planning to come home?[22]

Three months later Kobayashi wrote to Noguchi in the same vein, saying, "*Approximately* when are you coming home?"[23] They kept at Noguchi about coming home, about sending money, and about telling them of his marriage plans. To be sure, Noguchi had been somewhat less than consistent in all these matters, and he was able to make only a limited response. Oshika's pathetic appeal arrived in 1912 just as Noguchi returned from a successful lecture he had been invited to give in Ottawa, Canada, where he had been lodged in luxurious accommodations and treated with great respect. Clearly it was time for the truth, and he told them as much of it as he could. But he had never told them of his marriage, which had taken place almost a year before, and he still could not do so.

Noguchi wrote entirely of his work. Now claiming that he planned to return to Japan in June of the following year, he wrote: "I will tell you everything when I come back, but I would like you to know that the scope of my work needs the world as a stage. I must come back to the U.S. for the sake of the people in the world and for the honor of Japan, and by the urgent request of the United States." He explained at length that Japanese had no influence beyond their own shores because their work had not been good enough to justify their eminence. "When I see the many Japanese professors and scholars with big titles who come to the U.S. and Europe, they bow their heads low everywhere but cannot say a word. They may be somebody in Japan, but once they cross the ocean, none of them seems worthy of respect." Noguchi thought it was his mission to alter this situation and believed that he could do it. "If my work progresses at its present rate, I imagine it will not be difficult to win the prize which is given by the late Norwegian, Mr. Nobel, in his will."[24]

It took all these elaborate words of rationalization to tell them that he would support the family from his ample earnings—"I am sure that I will receive a better salary than the Japanese Prime Minister within a few years"—but he did not intend to return to Japan for more than a visit. But he still could not summon the courage to reveal that he had been married for almost a year to a woman whom he had never seen fit even to mention to them.

The widow of one of the members of the Rockefeller Institute (who is not willing to be quoted) remembers how pleased and relieved they all were when they learned that Noguchi had married. She said that on more than one occasion he had turned over his life insurance policy and made other extravagant promises to some casual female acquaintance whom he had met in the course of a festive evening, and the business manager of the institute had to visit the woman and negotiate Noguchi's release from his commitments.

Noguchi was married to Mary Dardis on April 10, 1911, but kept his marriage secret from the institute for some time. He and his wife moved into an apartment across the hall from Ichirō Hori, a Japanese painter whom Noguchi had met at the Nippon Club in New York a year or so before. Hori said that Noguchi had told him to keep the address secret.[25] We are told that Noguchi concealed his marriage because the retirement plan at the institute made a special provision for the widows of members, and Noguchi feared that the extra cost of this allowance for his wife might cause the institute to refuse him the further promotion to full membership, the one last step in his advancement to the top level.[26] There is no evidence that this fear on Noguchi's part was justified, but the circumstances of his marriage warranted some uneasiness about his wife's reception by Flexner.

The witnesses to the ceremony were Mr. and Mrs. Jacques Grunberg. According to Mr. Grunberg, a musician, when he was seventeen years old he was sent one evening to replace the pianist in a small orchestra at Luchow's Restaurant when the regular pianist fell ill. At that time

Mary Noguchi, Hideyo Noguchi Memorial Association

Luchow's was very popular with people in New York's theatrical and musical circles, and on this particular evening young Grunberg saw a young lady whom he described as the most beautiful he had ever seen. He managed an invitation to her table and was introduced to her. She took such a liking to him that she invited him to dinner at her apartment some days later. When he arrived, he found the lady (Myrtle is all of her name that we know), her roommate Mary Dardis, and "this little Jap, Noguchi." The two couples enjoyed their evening together and became a regular foursome.

On one such evening, after they had been drinking quite a bit and at a rather late hour, one of the group suggested that the two couples get married. They all thought this was a splendid idea and set out to find someone to perform the ceremonies. After wandering around for quite a while without success, they found themselves rather depressed in a bar in Hoboken, New Jersey. When they told their sad story to the bartender, he said he knew someone nearby who could solve their problem, and he dispatched a messenger to get the man out of bed. The two couples were married on the spot.

By the time Grunberg told this story in 1969, he and Myrtle had been divorced for many years and he had remarried. He had no knowledge of Myrtle's whereabouts or whether she was even still alive, and said that when they separated she had taken all of their letters from and souvenirs of Noguchi. The Grunbergs were still living together when Noguchi sailed for Africa in 1927, and the two couples had continued their friendship throughout the sixteen years between their joint wedding ceremony and Noguchi's departure.[27]

This story of Noguchi's marriage has never before been told in print. His wife is usually described as the daughter of a landlady, or as a maid in his boardinghouse, suggesting that she was not of the same social level as the wives of other members of the institute. We do not know how they met.[28]

What, then, do we know about the lady who became Mrs. Noguchi, but whom for the most part he kept carefully hidden? She and her three brothers were born in Scranton, Pennsylvania, to parents who had been born in Ireland. Her brothers were working in the Scranton coal mines as teenagers. She stated that she was thirty-five years old at the time of her marriage to Noguchi,[29] but although she had left home at least ten years before, we do not know where she was or what she was doing. According to Grunberg, she was associated with the theater, but was not a performer. He described her as a very large, strikingly beautiful woman, a description that accords with that of others. My father recalled with a chuckle, "Noguchi always said that the thing he liked best about the United States was that everything is so big, especially the women. And he got a big one!"[30]

Grunberg considered Noguchi's wife a heavy drinker; he said she was a sweet person when she was sober, but when she got drunk she could be pretty rough, even beating up Noguchi on such occasions.[31] Hori reported that Noguchi and his wife both liked to drink; they tried to stop drinking at the time of their marriage but failed. Noguchi would come home late after a long day, and would often find that his wife had been drinking to fight her boredom and that would start an argument.[32] She seems to have had little education and no understanding of his work.

Some of the nonprofessional staff at the institute came to know her somewhat better, and with greater sympathy, than the professional staff members. Bernard Lupinek, who started working at the institute in 1911 at the age of fourteen and retired from the institute in 1970, spoke of Noguchi's having been a kind of inspiration to him in his youth. Lupinek went through high school and college by taking night classes while he was working in the daytime in some menial capacity; he eventually became superintendent of buildings and grounds at the institute, spending his whole working life there. He knew every plant and brick on the place as well as every person who had ever passed through the institute's doors. In his early days, Lupinek slept at the institute as a sort of night watchman, and he remembered "dozens of times" when the night alarm would ring and he would open the door to admit Noguchi— at ten o'clock, midnight, or even two in the morning. Having climbed over the fence and rung the bell, Noguchi would tell Lupinek, "I couldn't sleep so I might as well be working." Lupinek had always assumed that these night visits took place before Noguchi's marriage and was surprised to learn that he was married in 1911.

Lupinek later often found it his job to conduct visitors to Noguchi's laboratory. There were many visitors, but he remembers Noguchi calling him — perhaps five or six times in all — to say that this wife was expected to arrive at the institute at a certain hour and to request that Lupinek meet her and conduct her to his laboratory. Lupinek always found her pleasant, and had the impression that she was trying to be "strict" with her husband. She wanted to slow down his pace of work. Maybe she was "bossy," but she was trying to take care of him, to get some order into his life. On most of these visits, Lupinek heard her try to persuade Noguchi to stop work for the day and go home. Lupinek, too, spoke of her as a handsome woman. He knew nothing of her drinking and would hear no word of criticism of her.[33]

Members of the professional staff at the institute insist that they knew nothing about Mrs. Noguchi, but they say it with an air that implies that they knew a great deal but were above repeating such malicious gossip. They did not know her—that is the real truth. As soon as the word was out that Noguchi was married, the institute held a reception at which the members met her as she stood in the receiving line, but they never had occasion to speak with her again.

The gossip that went around the institute was probably the usual combination of truth and distortion. Rockefeller totally abstained from alcohol of any kind, and not a drop of liquor was ever served in the institute dining rooms or at any institute function. It would, therefore, have been desirable to keep any stories of drinking by any of the staff members or their families as quiet as possible. Mrs. Noguchi's lack of education, her connection with the theater (such people were considered "fast"), and probably even the fact that she, a Caucasian woman, had married an Oriental man, all would have denigrated her even in the eyes of such intellectual people who considered themselves above snobbery.

Noguchi understood these things. He believed that this marriage could hurt him at this critical time in his professional advancement. This belief must have been the reason for his secrecy, and yet he needed a loving wife and a stable home. The fact that he married her, even with the help of a little alcohol, and in spite of the threat to the advancement that meant so much to him, must be a measure of his loneliness and of some caring quality he found in her. How Noguchi planned to fit her into his life in Japan, if he returned, is something he never explained. Mary Dardis Noguchi had surely been acquainted with poverty in her childhood as had many people in Scranton in those years. At the time of her marriage, she probably knew that her husband's family were poor farmers in Japan, but neither she nor anyone else in this country could envision the true level of the poverty from which he had come and from which his family was not yet securely free. Even he postponed facing it as long as he could.

12

The Noguchi Mystique

BETWEEN 1906 AND 1913, NOGUCHI'S EARLY WORK ON SNAKE VENOMS WAS published in a lavishly produced volume, and his ambitious studies on syphilis captured the attention of the scientific establishment. Already in this work, one can see the genesis of his brilliant achievements as well as the personal limitations that would one day cause that brilliance to be obscured by episodes of conspicuous misjudgment.

Although Noguchi kept postponing the enormous task of writing the monograph on snake venoms for the Carnegie Institution, he managed to finish it some time in 1908, and it was published in 1909. Noguchi stated in his preface: "No single work in the English language exists at this time which treats of the facts of zoological, anatomical, physiological, and pathological features of venomous snakes, with particular reference to the properties of their venoms."[1]

Clark refers to the monograph as a "distinguished publication with many drawings by Noguchi."[2] The Carnegie Institution subsidy included a special allowance for the thirty-four plates, six of which were reproductions of Noguchi's own elegant drawings, and the several photomicrographs, which he may also have prepared.[3]

The English is obviously Noguchi's; there is enough editing to correct the grammar without eliminating his inimitable style. The literature in the field is reviewed and all contributors are recognized. A monumental undertaking, there appears to be an especially detailed presentation of Mitchell's work, suggesting that the monograph was a tribute to this pioneer and benefactor. Flexner's and Noguchi's work, presented in such a context even though it had been published elsewhere, would surely have drawn the respectful attention of a small but select group of readers. The monograph is difficult to find today; aside from the Library of Congress and the Carnegie Institution, it exists mainly in private collections.

During the time Noguchi was working on the monograph, he started his work in syphilis. In 1905, Schaudinn and Hoffmann had identified the spirochete that causes syphilis; in 1906, Jacob Wassermann developed a serologic test for use in diagnosis of syphilis; and, in 1907, Erhlich synthesized the first arsenic derivative (Arsephenamine or Salvarsan), which was the only treatment for syphilis for many years. Almost from the beginning, Noguchi had a threefold purpose: first, he sought a practical refinement of the Wasserman test to make it simpler and more accurate; second, he sought to grow the spirochete *Treponema pallidum** in pure culture; and third, he attempted to find the organism in pathologic preparations from brains and spinal cords of patients who had died of *tabes dorsalis* or of paresis. This third objective, supported by transmission of syphilis from tissues of such patients to susceptible animals, would establish that paresis and *tabes dorsalis* were later manifestations of syphilitic infection.

It became the common belief that Noguchi had accomplished all three of these objectives, and he certainly believed it himself. He developed, and revised many times, what came to be known as Noguchi's butyric acid test. He believed that he had grown a pure culture of the organism, but his results were difficult to repeat and a pure culture has not been achieved to this day. But he did find the syphilis organism in the tissues of paretic and *tabes dorsalis* patients, an accomplishment that not only brought him recognition but added to the general confidence in both of his other claims. Since Noguchi worked also on trachoma during this period, and further announced the discovery of the causative organisms of rabies and poliomyelitis, and since he appeared to have convinced some informed and astute people, it began to look as though he really had some unique talents.

In his own description of his progress for Chiwaki, Noguchi emphasized the attention his work was receiving in Europe and even in South America. He told of doctors coming to his laboratory to learn his methods and of being invited to give lectures, and he announced a new objective:

> I am sending for your laughing inspection four papers about this discovery. After reading them, will you please submit them to the Ministry of Education. I expect to publish another one during the month of May and will send this to you as well. As I told you, Dr. Kitasato represented Japan at the International Congress on Tuber-

---

*In 1905, Schaudinn replaced the generic name *Spirochaeta* with the new term *Treponema*. This classification was not generally accepted for several years, but Noguchi adopted it relatively early, and we use it except in quotations using *Spirochaeta*.

culosis last year. A reprint of the paper which I presented at that time will be sent to you some time soon.[4]

Noguchi described the snake venom monograph and said he would send a copy of that as well for Chiwaki to submit to the Japanese Ministry of Education, all in support of his application for a degree from a Japanese university.

Noguchi's determination to design a single diagnostic test for syphilis, which would be an improvement on the Wassermann reaction test, was a precursor in his mind to the eventual development of a syphilis anti-serum for prevention and treatment. It became his objective here, as it did with almost every disease on which Noguchi worked; he viewed it as the "crucial point" of his research with snake venoms. Practical application of this work in syphilis might be even more lucrative than the rattlesnake antivenom that he had wanted to produce in Philadelphia. Many workers tried to develop a diagnostic test for syphilis, Noguchi being only one of scores who failed in that task as explained by recent writers on the subject:

> Human infection with *Treponema pallidum* stimulates the host's defense mechanisms and provokes a complex antibody response. The detection of one or another of these antibodies is the basis of serologic testing. The ideal single test would be easily and quickly performed, it would be highly sensitive and specific, and the results would be readily reproducible in different laboratories. At present, no such ideal test is available. Although over 200 tests for syphilis have been described, only a few are now used. [1968] By carefully selecting and familiarizing himself with two or three tests, the clinician may diagnose syphilis in nearly any stage using test currently available. . . . Serologic tests are a definite aid in the diagnosis of syphilis at any stage, and they are the basis for the diagnosis of latent syphilis.[5]

Obviously no one anticipated the enormous difficulties that this problem presented. Noguchi clung to the problem for years, and he was drawn to it as a problem for which his technical experience had prepared him well. It suited his own ambitions in its practical importance and in its potential for creating goodwill for the institute and Rockefeller. He knew that the Japanese would quickly recognize the value of such a test.

Among the physicians who came to work in Noguchi's laboratory was a urologist, David Kaliski, a young man starting practice in New York City, who was interested in learning the techniques of the Wassermann reaction as well as in Noguchi's modifications. Dr. Kaliski worked as a volunteer and helped in the writing of the first edition of Noguchi's *Serum Diagnosis of Syphilis,* which was published in 1910.[6] As a urologist, Kaliski maintained his interest in the diagnostic aspects of Noguchi's work long after his volunteer work at the institute stopped. Noguchi and Kaliski published together as late as 1918,[7] and whereas we know of social and other professional contacts between them, records are not available at this time.[8]

Flexner wrote of Noguchi in his diary: "Once he was started on a problem he would pursue it to the bitter end. This habit of his led him to go on with the Wassermann reaction long after he had exhausted the problem for himself. But it led to the cultivation of the spirochaetes which proved highly rewarding."[9] Corner writes that Flexner did not approve of Noguchi's working on the "highly technical problems of complement-fixation and the immunology of syphilis," but rather encouraged his attempts to cultivate the spirochete after the initial success in confirming its discovery.[10] Others have reported that Flexner was constantly trying to get him away from this work only to find that Noguchi was again trying to improve the Wassermann reaction test.[11] However, Noguchi was promoted to the rank of associate at the Rockefeller Institute in 1907, and to associate-member in 1909, so he had every reason to believe that his work was approved.[12]

The second path of Noguchi's interest in syphilis was the growth of the organism in pure culture, a necessary step toward the establishment of the causal relationship as defined in Koch's postulates.[13] It was also a technical requirement for making syphilis immune serum. When Noguchi believed that he had grown a pure culture for the first time in February 1911, he wrote to Kobayashi in joyous excitement. "I feel as if I am dancing in heaven. . . .It is strange that success has become mine while all the other famous pathologists have been working without success. . . . This success in cultivation in vitro could greatly influence the prevention and treatment of syphilis in the future. These also are in my hands."[14]

Corner's description of Noguchi's style of work in his attempt to culture the organism portrays Noguchi's laboratory as his contemporaries saw it. "The syphilitic poison will. . .grow and multiply in a rabbit's testicle while the contaminating bacteria largely die out, leaving a more or less pure culture of spirochetes. Noguchi followed this up in characteristically hectic fashion, inoculating rabbits with ten different human strains, setting up hundreds of culture tubes filled with many kinds of nutritive media, sampling the cultures on thousands of microscopic slides."[15] Observers often questioned his system, or lack of system, for labeling his tubes, but Noguchi insisted that he had a system and knew what each tube contained. From this exuberant beginning, Noguchi devised an ingenious apparatus for further isolating the spirochetes from the contaminating bacteria, and wrote a convincing paper claiming that the end product of his procedure was a pure culture of the spirochete.[16]

No questions were raised for some time, but as others continued to have difficulty in repeating his results, the problems were extensively worked over and discussed in the literature. Hans Zinsser, a young internist in practice in New York City, came to work as a volunteer in Noguchi's laboratory to learn the techniques for serologic tests. Zinsser's influence extended, however, to Noguchi's own efforts to grow the syphilis organism and to the evaluation of those efforts by others. Zinsser turned later to teaching and medical research and became the original author and editor (1910) of *A Textbook of Bacteriology,* which has gone through many revisions, first by Zinsser and more recently by others.[17] It has remained a standard work on the subject for medical students to this day. Zinsser also wrote an autobiography in which he spoke of Noguchi and of the general atmosphere of medical research of the period with humor and sympathy as well as scientific insight.[18]

Zinsser has been described as an ebullient, friendly man "with bands on all four corners;"[19] he, in turn, describes working with Noguchi with trust and affection: "For the time being, I was engaged in introducing the so-called 'Wassermann reaction' into the laboratory practice of St. Luke's Hospital. . . . This reaction. . .was taught to me, out of the kindness of his heart, by my friend Noguchi, in whose room at The Rockefeller Institute I sat for hours to pick up what I could."[20]

Zinsser was one of a few who reported that he had repeated Noguchi's culture work, even devising slight improvements in technique.[21] Zinsser continued to report successful cultivation of the syphilis spirochete until 1934 when he accepted the fact that his own cultures were avirulent and wrote, "It is our opinion that virulent *Trepnoma pallidum* has not been cultivated in artificial media."[22]

In 1966, under the auspices of the World Health Organization, R. R. Willcox and T. Guthe published a bibliographical review of the enormous body of material that has accumulated from years of effort by many workers to cultivate this organism. The introduction to this work states, "The intimate nature of the organism, and the key to its successful cultivation outside the bodies of men or animals (i.e., virulence) still remains to be determined." And in a historical survey, they continue: "Period 1911-20. . . .Dominant at this period was Noguchi who, in 1913, demonstrated *T. pallidum* in the brains of sufferers from paresis. . . .Noguchi was the first to claim that the *T. pallidum* of Schaudinn had been 'proved beyond all doubt' to have been obtained in pure culture. . . .Virulence, however, was not maintained."[23]

But these evaluations came later. Shortly before his death, when Zinsser and others still believed that they had been able to confirm his culture work, Noguchi wrote: "The virulence of the organism is lost very early in the course of cultivation, and pathogenicity tests to determine whether a strain is actually *T. pallidum* must be made soon after isolation."[24] He had never doubted his conclusions and he died believing in them. He never learned that such statements as "proved beyond all doubt" would foster suspicion and antagonism when other events came along to trigger such reactions.

The discovery of *T. pallidum* in the tissues of paretic patients turned out to be the most stable achievement of this period of Noguchi's career. At the time it had additional merit as a discovery that drew attention in a field far removed from bacteriology. In a discussion of the history of American psychiatry, John C. Whiteborn wrote: "In the organicist tradition, the outstanding psychiatric achievement as well as the final and conclusive link in the demonstration of the etiologic role of syphilis in general paresis was Noguchi and Moore's demonstration of the spirochete in the brains of general paretics."[25] Beyond this, Noguchi's discovery established for the first time that a psychosis could be caused by an organic agent.

The New York State Pathologic Institute was located at Wards Island State Hospital, a hospital for psychiatric patients on an island in the East River almost directly opposite the Rockefeller Institute. Two early staff members at the Rockefeller Institute, Phoebus Levene and James B. Murphy, had worked at the Pathologic Institute and were well aware of the problems that concerned the psychiatrists. About 20 percent of the first admissions to the New York State hospitals for the mentally ill were patients suffering from paresis, a psychotic disease that inevitably ran a chronic course to the patient's death within five to seven years. When Noguchi confirmed the Schaudinn-Hoffmann discovery of the spirochete, and especially when he announced the pure culture of the organism, psychiatrists looked to him for further help in determining the relationship of syphilis to paresis and *tabes dorsalis*.

In 1910, my father, who was clinical director of Kings Park State Hospital on Long Island, acknowledged Noguchi's help in setting up laboratories for clinical examination of cerebrospinal fluid as well as for research at that hospital. He reported using Noguchi's butyric acid reaction test and finding it more sensitive than the Wassermann reaction for spinal fluid, adding, "Noguchi. . .had prepared for us all the antigen and ambocepter tests that we used. He also spent about two weeks at our laboratory and helped us materially by making many of the tests."[26] In 1913, in collaboration with J. W. Moore, a psychiatrist at Wards Island, Noguchi reported finding *T. pallidum* in slides from the brains of paretic patients.[27] The popular story of his poring over slides on which others (including Moore) had been unable to find spirochetes, and finding them in the late hours of the night, drew as much attention to Noguchi's intense manner of working as to the discovery itself.

His friend Ichirō Hori, who still lived in the apartment across the hall from Noguchi, told of Noguchi's bursting in on him in the middle of the night. Dressed only in his underwear, Noguchi was shouting, "I found it! I found it!" as he danced all over the apartment.[28] Noguchi then dressed and went to Flexner's apartment, and Corner picks up the story:

Noguchi brought to the quest no new method, but only his own high-strung determination, persistence, and visual acuity. Collecting 200 brains from cases of general paresis and twelve tabetic spinal cords, he made innumerable sections—staining them by various methods in batches of 200 each—and tortured his eyes through long nights at the microscope, intently looking for the tiny spiral threads that could too easily hide themselves among the interwoven fibers of the brain. Working at home all one night in 1912, he at last, as dawn approached, came upon the organisms in one of his slides, sparsely scattered through the substance of the brain. Greatly excited, he left his microscope on the dining table and hurried through the streets to call Flexner out of bed at 5 a.m. Coming with amused tranquility, Flexner shared Noguchi's excitement when he too saw the spirochetes. Once Noguchi knew where to look in the paretic brains, he found the spirochetes quite readily again and again; in the tabetic spinal cords they were also found, though never easy to see. Thus Noguchi had proved conclusively that general paresis and tabes dorsalis are indeed late stages of tertiary syphilis of the brain and spinal cords respectively. For this achievement the Association of American Physicians gave him in 1925 the first award of its prized Kober Medal.[29]

Noguchi spent considerable time and energy in an effort to perfect the substance he called "luetin" for the single test of syphilis, but in part because he never had a reliable supply of the organism in pure culture from which he could make this substance, this effort never yielded practical results. However, the name was trademarked in the United States as well as in several other countries, and Noguchi continued to hope that techniques for the large-scale production of this substance would materialize.

In the period around 1910, Flexner's public statements in praise of Noguchi's achievements and talents began to be couched in language that generated even more resentment among scientists that Noguchi's own words, "proved beyond all doubt." Such statements by Flexner reached their peak of arrogance in the obituary that he wrote after Noguchi's death, but this encomium was representative of the attitude he had repeatedly expressed from the time of Noguchi's supposed cultivation of the syphilis organism.

The culture of the syphilis spiral was made to yield luetin, a soluble extract based on tuberculin, of use in detecting latent and congenital syphilis. There is no better incident than this to bring out Noguchi's faultless and infinitely varied technical skill. The culture medium we are considering is not only very variable in itself, because of the chemical complexity of the materials entering into its composition, and therefore exceedingly difficult to keep approximately constant, but it demands constant modification in order to adapt it to the many organisms the cultivation of which he accomplished through its use. It is no wonder, therefore, that so many of Noguchi's would-be followers have failed in their efforts. . . . With factors so variable in their nature, what he perhaps did not do, and what such consummate masters of technique almost never find it possible to do, is to put into words those subtle, imponderable yet essential twists and turns of method used by them, often unconsciously, in adapting a medium to a recalcitrant microorganism. The patient and resourceful among bacteriologists have learned in time to repeat what Noguchi has done, but the mass of the conventional among them undoubtedly soon tired and gave up the unequal contest.[30]

Such an attitude actually encouraged the very kind of "ambition" that Flexner came to criticize in Noguchi. Beyond that, however, the glamor of the Rockefeller Institute, the real skill, ingenuity, and hard work that Noguchi put into his work, plus a special aura of the unknown by reason of his being Japanese — all these supported Flexner's picture of Noguchi as the master of technique, the man who could do things that others found difficult or impossible to repeat. It was a picture that Noguchi was all too willing to have painted of him, and he was undoubtedly the most seriously deluded believer in the mystique that started to grow around him at the time of his experiments in the cultivation of the syphilis organism. It gave him the feeling, which he expressed to Kobayashi after his successful lecture in Ottawa in 1912, that his work needed the world as a stage; and it also gave him the courage to tell his family that he did not plan to return to Japan.

In a letter to Flexner written during the summer of 1910 Noguchi added almost as an afterthought to the main subject: "I am sending you herein a manuscript of one of the works done before the vacation. Bronfenbrenner has assisted in this work in part. Of course, any part of work done by him was under my close supervision and much of it was repeated by me."[31] With the exception of one other casual mention of J. J. Bronfenbrenner, this is the only record of Noguchi's speaking of him, in spite of the fact that they published ten papers together. The work reported in these papers was of a biochemical nature, a continuation of the immunologic studies that Flexner had been trying to discourage. Noguchi's ambitions were far too pretentious for the knowledge of biochemistry of the time, and he was not the only one who floundered in the confusion of his ignorance. Although Bronfenbrenner is said to have become very critical of Noguchi and his methods of working, the critical words that are attributed to him do not always appear to be related to the work of these papers.

Bronfenbrenner was a graduate student at Columbia during much of the time he worked with Noguchi. He was an orderly worker, and if he tolerated Noguchi's disorder only to find out later that the work was less reliable than he had thought at the time, one can see that he might have become angry. But there are reasons to suggest that Noguchi may have had loosely defined objectives in his mind that were not always obvious to Bronfenbrenner.

Several people have given their recollections of Bronfenbrenner's experience with Noguchi. His son writes:

> As luck would have it, I myself became interested in things Japanese during World War II. I served as a Japanese language officer in the Navy, have been back and forth to Japan seven or eight times, and have married a Japanese girl. In connection with this marriage, my father expressed the view, apparently gathered from Dr. Noguchi, that most civilized Japanese left the country as soon as they could and that those remaining in Japan voluntarily were mostly "barbarians." If this represents Dr. Noguchi's position, it is very different from that ordinarily attributed to him in Japanese school textbooks.[32]

Bronfenbrenner left the Rockefeller Institute in 1913, returned in 1923, and finally left again in 1928 to become professor of bacteriology at Washington University in St. Louis. A young woman who worked at the Rockefeller Institute as a technician in 1927 and 1928 went on to become a graduate student working with Bronfenbrenner at St. Louis. Although she never knew Noguchi in New York, she knew the atmosphere of the institute, which she describes as "stuffy." A number of people went to St. Louis from the institute at the same time as Bronfenbrenner, all of whom had been of much higher rank at the institute than she. As a lowly technician, she had had no contact with any of these people in New York; she had seen them but says that the atmosphere of the institute was so caste-oriented that no one spoke to anyone outside of his or her own bailiwick. In St. Louis they were much less formal and soon were recalling in a friendly manner the fact that they had seen her at the institute.

She describes Bronfenbrenner's attitude as, "In general, pay no attention. That's Noguchi." Speaking of someone whom he wished to criticize, he would say, "Oh, he's another Noguchi!" She believes that Bronfenbrenner was a "real scientist," and although he would not have said that Noguchi was dishonest, Bronfenbrenner felt that his work was unscientific. She goes on to say: "I remember one incident, either with treponema or leptospira, when Dr. Bronfenbrenner said, 'Noguchi had us sitting there drawing pictures of everything we saw [under the microscope] of the treponema. We had to sit all day, drawing pictures, pictures. . .and then he'd go home and he'd arrange them in a way that would make a nice life cycle. He was the God-almighty of the way treponema was put together from the pictures we drew.' And I mean that he said that in the most scornful way."[33]

Nevertheless, right after he left the institute in 1913, Bronfenbrenner continued working with some of the techniques and problems he had begun in his work with Noguchi, referring throughout one small paper to the effi-

ciency of Noguchi's method and to the various details that were maintained in his work as Noguchi had designed them.[34] In spite of his scornful
remarks at a later date, in the days of his association with Noguchi
Bronfenbrenner seems to have been trusted to hear such expressions of
Noguchi's hostility toward Japanese people as no other Westerner has
reported. This was the time at which Noguchi was negotiating for his
degrees from Japanese universities. The tone of his critical remarks to
Bronfenbrenner echoes the tone of the tirades he wrote to Kobayashi. Was
there a friendship between Noguchi and Bronfenbrenner that permitted
these confidences at one time, but which turned sour later? Or did
Noguchi fail to realize that the student, Bronfenbrenner, held him in such
contempt? The "pictures, pictures" that he had his workers drawing had
no relationship to immunologic or cultivation work which they published.
The drawings did contribute to Noguchi's understanding of the structure
and motility of many forms of spirochetes and, in spite of
Bronfenbrenner's contempt, the deductions Noguchi made in these matters were later found to be entirely correct.[35] Noguchi needed a student
who could take the formal courses and provide the background in
chemistry and other basic material that he lacked. If he was trying to
establish a collaboration with Bronfenbrenner that might continue into
the future, his overtures to Bronfenbrenner were either not recognized or
were rejected.

From 1924 to 1926, after Bronfenbrenner returned to the institute as a
full staff member, Philip Reichert worked for him as a postdoctoral assistant. Dr. Reichert's recollections of both Noguchi and Bronfenbrenner
provide still another view of these men:

> I have many reasons to remember Noguchi with affection. As soon as
> anyone shows an inclination to be helpful to young people he incurs
> the penalty of being constantly bothered. But Noguchi never lost his pa-
> tient smile, his willingness to listen, his delicate compliment of telling
> you that you were doing something important and doing it well. I need-
> ed that.
>
> My room was. . .directly above Noguchi's lab and just down the cor-
> ridor from Simon Flexner's office. My immediate superior was Bronfen-
> brenner who specialized in sarcasm; he seemed to feel that was the best
> way to get work done, and he drove us hard. . . .I never tried to ask him
> questions. He gave me the feeling that he was surprised that I could ask
> such simple things. . . .But Noguchi was always there, always stopped
> what he was doing, always gave you the impression that he was
> delighted to see you, and how wonderful it was that you thought he
> could help you. . . .Noguchi was an outstanding technician. He usually
> kept his crippled hand in his pocket, but if you surprised him at his
> bench he would be holding more test tubes in that pathetic hand than
> you could count, deftly flaming the top of each, seeding, stopping with
> cotton. . .so quickly it was a joy to watch. He loved technical things and
> since I was a gadget maker he was always interested in my 'imventions'. . . .

He would look at the gadget with wonder in his eyes, tell me how clever I was, ask me the details of how I managed to think up something so useful, and generally make it a tough job to get my hat back on my head. No wonder I loved him. In the atmosphere that was charged with tension and frustration that ego salve was merciful.[36]

The interpretations suggested by these selected examples of personal relationships among these men and women are certainly no more than speculations. We really know only that people saw each other differently even as they had differing views on the institute itself. Noguchi needed help and he knew it; yet he could not ask for it and he was never an easy man to guide or control. In the Philadelphia days, when he and Flexner were both learning immunology through his supervised projects, this had been less of a problem. But in the period between 1906 and 1913, the situation had entirely changed.

Supposedly qualified to work with some degree of independence, Noguchi's ambitions were even further stimulated by the highly competitive climate created by so many ambitious men. His enormous volume of ingenious work, leading to valuable techniques and claims that he had solved obviously important problems, led also to his gradual appearance on the international scene as a kind of scientific wizard. He welcomed, and was even further goaded by, this reputation just as the institute welcomed and nurtured it. The chaos of his personal life, exacerbated by the constant reminders of the misery of his family in Japan and his own disorganized promises to help them, not only reflected the anxiety in his own ambitions but contributed further to that anxiety.

The pattern that led to his destruction was firmly established during these years: a self-perpetuating cycle of squandering his personal gifts and his magnificent opportunity in frantic efforts to do too much too fast. With so much of the focus on the dramatic, on problems that were readily understood by the general public, Noguchi gained recognition and adulation such as he had not even known enough to dream of. Thus, the words he used, "proved beyond all doubt," to report that he had grown *T. pallidum* in pure culture eventually singled him out for failure to substantiate claims, whereas dozens of other workers who have met with similar frustrations in their efforts to cultivate this organism were subject to no such disparagement.

13

The Critical Year

It is only by risking our persons from one hour to another that we live at all. And often enough our faith beforehand in an uncertified result is the only thing that makes the result come true.

— William James, *The Will to Believe*[1]

NOGUCHI FOLLOWED HIS REPORT OF MIDNIGHT DISCOVERY OF *T. pallidum* in the tissues of paretic patients with a note on the transmission of some sort of disease from tissues of one case of paresis (out of six) to one rabbit (out of thirty-six) and his recovery of the spirochete from that one rabbit.[2] Noguchi said that a full report would be forthcoming but it never appeared. He was on his way and it did not seem to matter; he was involved with so much more.

In December 1908, in Vienna, Landsteiner and Popper had announced the isolation of the poliomyelitis virus. Dr. Flexner had taken up this problem and assigned various parts of it to several young men on his staff. Flexner and his assistants were able to transmit the disease to monkeys, and, since Noguchi was thought to have been successful in the cultivation of *T. pallidum* "by special methods," Flexner assigned him to cultivate the polio organism.

An important feature of Noguchi's cultivation method was his placing of small bits of living tissue in the culture. Theobald Smith, who had devised this procedure to remove oxygen from cultures of organisms that require an oxygen-free environment, suggested it to Noguchi.[3] Since Flexner and Noguchi had some success in producing the disease in monkeys with Noguchi's cultures, they both thought that an organism which appeared regularly in their cultures was the poliomyelitis organism. Although they spoke of viruses, it was not known at that time that a virus is a parasite on living tissue and requires such tissue for its growth and survival. Clark, who worked at the institute on this poliomyelitis project, remarks that it seems highly probable that Noguchi and Flexner had some multiplication of the virus in their media. "There is no doubt that the experimental disease produced from the cultured material was poliomyelitis."[4] In their haste and ignorance, Noguchi and Flexner did not consider that their organism might be a contaminant while the virus was there but too tiny to be visualized under their microscopes.

Almost twenty years later, Flexner, still believing, wrote in his diary: "Noguchi did the culture work but became hopelessly entangled. He was confident that the organism was a protozoan with a varied morphological life cycle.* I had put N on the cultivation and followed the steps and rescued finally the globoid bodies which I named, and I made the inoculation tests. We published the work jointly but circumstances later revealed N's disappointment."[5]

"N's disappointment," which was not related to the work but to the sharing of credit, became widely known and even reached Rockefeller's office. Jerome Greene had been business manager of the Rockefeller Institute until 1912 when he moved on to join the personal staff of John D. Rockefeller, Jr., where "he had a distinguished career as banker, internationalist, and trustee of practically all the Rockefeller philanthropic boards."[6] The son of a missionary, Greene was born and lived through his childhood in Japan. He spoke Japanese, and was always sympathetic to Noguchi and affectionate toward things Japanese. On May 20, 1913, Greene wrote Dr. Welch, who had been chairman of the institute's Board of Scientific Directors since the founding of the institute, that he was bringing to Welch's attention for the second time the persistent rumors that Noguchi had not been given proper credit for the identification of the poliomyelitis organism. Greene believed that "this feeling, however unwarranted, is capable of doing an injury to the Institute. I dread the consequences of a general belief that Dr. Flexner was inconsiderate."[7] When the matter was brought to his attention, Flexner responded with a detailed report of the manner in which the work had been reported both in public statement and in print, and with elaborate assurances that no such discrimination had been intended in the past and would not occur in the future.[8]

During the spring of 1913, Noguchi received an invitation to lecture as a guest at the annual meeting of the Association of German Naturalists and Physicians, which was to be held in Vienna in September. This was a step

---

*The fundamental difference (in subcellular organization) between bacteria and protozoa was not known in Noguchi's time.

into the inner circle of the world of medical research, and Noguchi turned his full attention and energy to the preparation of the kind of material that would guarantee his acceptance by these arbiters of medical science. He planned to demonstrate the evidence for the claims he had made in the cultivation of various spirochetes, particularly the agent causing syphilis, and the evidence of the association of syphilis with paresis. His claims had been questioned in Germany; they were so arrogant, these Germans, especially about anything that came from America. Noguchi knew, however, that the Germans held the key to the door of scientific approval not only in America but in Japan as well. So he planned to overwhelm them with the quantity as well as the quality of his achievements.

The presumed cultivation of the poliomyelitis organism would be added, but it carried the burden (in Noguchi's view) of having been announced in collaboration with Flexner. Among the various spirochetes he had cultivated and described in publication was the organism that causes one form of relapsing fever, but that did not seem to have created much interest. It was just another disease that was not a great threat either in Western Europe or in the United States.

Most of the institute's top staff left for the summer months, but Noguchi could not fritter away valuable days in vacation. However, he had to implement his plans through correspondence, and it is to this circumstance that we are indebted for at least a partial view of his power to deceive himself and to coax Flexner into accepting material for publication against the latter's better judgment.

Alexis Carrel, a colleague at the Rockefeller Institute who had received the Nobel Prize in 1912, was in France for the summer of 1913. Noguchi started his campaign by writing Carrel, "I have just completed my cultivation work of rabies and am writing an article. I am now *absolutely* sure of my work. . . .If opportunity is given I would like to demonstrate my work (projection) in Paris while in Europe in October. . . .Where should I speak?"[9]

Five days later, on July 30, Noguchi wrote Flexner in New Hampshire:

At last I managed to present to you my results of cultivation of the rabies virus. As the results so far obtained are so definite and the time at my disposal before my departure for Vienna (will sail at the end of August) is rather short I decided to write out my work up-to-date in form of a manuscript rather than to tell you these facts in fragments. As you will notice from the text and the accompanying illustrations I have succeed to grow the Negri bodies, as well as the granular and pleomorphic bodies from various strains of viruses. I can produce rabies with the cultures of any of the forms described at any time. These forms are all related to each other and of course cannot be grown in any other media.[10]

Noguchi described the lucky accident that had brought these bodies to his attention and another accident that "destroyed some of the best preparations!!! It was a sad thing, but I have saved *a few* fragments of slides showing the bodies. . . .I am *extremely anxious* to hear from you about the work. . . .Of course you will decide in regard to the publication, but I would like to express my wish for a prompt appearance, since, as you know, the work may be duplicated by outsiders and lose its freshness."[11] Noguchi asked whether he might see Flexner in New York before he sailed to Europe, and Flexner replied at once, telling Noguchi to bring his slides and all relevant material to New Hampshire where they could discuss the publication.[12]

Noguchi visited Flexner, who had considerable doubt and instructed Noguchi to stop in Boston on his way back to New York and show the work to Theobald Smith at Harvard. Arriving in New York, after a long day in which he had seen Smith and then taken the long train trip home, Noguchi wrote Flexner. He said that he had shown all his evidence to Smith and described Smith's reaction to it: "After looking them over he sat down and thought for some time. He then looked up some books and his own memoranda on certain protozoa. He then told me that he has never seen anything like these nucleated forms. He said that these bodies are actively multiplying and not artifacts of any kind." Smith compared Noguchi's slides with some old rabies slides in his own collection which had been prepared with different stains but "he was unable to decide about the bodies in the culture whether these are the cultivated forms of the Negri bodies or not. . . .He advised me to continue the work further to clear up the findings. You will see that it is too new and striking for him to express any definite opinion beyond that. . . .The best person who would be able to decide the situation will be the man who has been working constantly with the problem. *I believe that you are convinced** of the similarity of the bodies found in the films from street virus (dogs) and in the cultures. Now the article is so modified that there can be no blunder even it is published (sic) at this stage of work."[13]

When he arrived at the institute the next morning, Noguchi found what must have been an enthusiastic cable from Carrel. He responded by writing of his visits to Flexner and Smith: "My demonstrations were satisfactory to him [Flexner] and also to Prof. Smith. . .and I am now authorized to publish a part of the complete work." From this distortion, Noguchi went on to ask Carrel to present a brief illustrated report, which he was sending, to the French Société de Biologie.[14] Shortly thereafter, Flexner wrote to Noguchi from New Hampshire:

---

*Italics mine.

I telegraphed Miss Butler last night to say that the brief paper on rabies
was to go into the September number of the *Journal*. I rather thought
that Dr. Smith would take the position that he did, namely, that is is
desirable to follow the matter of the cultivation up much more closely
before reaching a final decision. On the other hand I think if we make a
careful publication at the moment, without committing ourselves too
much, that it will be safe, and on that account I have asked Miss Butler
to go ahead, but to send me a carbon copy of the paper so that I can
assure myself that the matter is cautiously presented. When you return
from abroad it will be necessary for you to devote yourself to a confirmation and extension of the work in every way that you can. It seems to me
that the finding opens up an interesting and important field, but just
how far it will go of course cannot be predicted now. As it is, I am very
happy that you have made this advance. Remember that I should like
you to send me right away some of the photographs of which you spoke,
showing the characteristic bodies, that I may have them before me when
I read the full paper. . . .Please see that this is done promptly.[15]

After telling Flexner how happy he was with this decision and complying
with the request for photographs, Noguchi appears to have had his own
measure of assurance from this negotiation. "Now all is over. . . .As soon
as I return from my trip abroad I will resume the work." But he made it
clear that in doing so, he intended to explore the biological characteristics
and immunity problems with the organism rather than the confirmation of
the work which was being so tentatively presented.[16] Noguchi wrote Flexner from the ship on the day of his sailing for Europe: "Of course the work
will be resumed upon my return, but you know that this is to convince
others who are skeptical."[17] The work was never resumed. Flexner wrote
later: "I tried similarly to forestall the publication of the life cycle of rabies
virus, but unsuccessfully. I induced N to confirm his own work on rabies
which he failed to accomplish. The matter was dropped, perhaps as well,
as it had become discredited in any case."[18]

Neither Noguchi nor Flexner was in New York when the brief note appeared in the *Journal of Experimental Medicine*, but it caught the attention of at least one journalist. Corner writes that "tyros" of the press often
sensationalized items which they picked up by reading this journal, finding
Noguchi and Carrel particularly good subjects for this kind of presentation.[19] In some defense of the press, it must be added that Flexner was
wary of journalists and reluctant to make any information available to
them. In this instance, on September 7, 1913, the *New York Times* (in a
front-page article) and the *Philadelphia Public Ledger* both printed
almost the full text of Noguchi's note as well as an identical, lengthy quota-
tion from Dr. George Gibier Rambeau, director of the New York Pasteur
Institute, and the same biographical sketch of Noguchi. Headlines and an
introductory paragraph in each paper asserted that Noguchi had isolated

and cultivated the rabies organism after other workers had sought it in vain for thirty years.

Dr. Rambeau is reported to have said: "If this report had emanated from any source but Dr. Noguchi I would be inclined to discredit it. If Dr. Noguchi states that he had successfully cultivated the parasite of rabies, it is doubtless true." Noguchi is not quoted, however, as having made that statement. If the journalist drew his sensational conclusion from the conservatively worded note in the journal, he could well have sensed that he was not in error as to Noguchi's opinion, and he certainly knew that Flexner approved whatever went into that journal. But with no further clarification of the background to tell us how this announcement happened to be made in this way, we are left to follow Noguchi through Europe.

Descriptions of his trip are found in his own letters and in reports of other participants in the congress. There is no doubt that Noguchi's performance astounded the scientists of Europe. The list of all the places at which Noguchi was invited to speak, and the distinguished people who entertained him, leaves out no place or person of importance.[20] From Paris he wrote to Flexner:

> At the request of Prof. Metchnikoff, I have made the dark field demonstration of my pure cultures of pallida refringens, micro and macrodentiums, calligyrum, etc.* It was a great triumph for me, *because I convinced them all of everything I have claimed**. . .*Prof. Metchnikoff wanted to see the rabies culture. This I showed. He was much interested and said I have the virus grown. Everyone who saw the presentations said they were convinced of the identity of the cultivated forms and Negri bodies. I spent several hours with them on two visits I made.[21]

Flexner responded with pleasure but again expressed the hope that the work was all right.[22] It is impossible to sort out Noguchi's optimistic estimate of the appraisal of his work by such a distinguished and widely known figure as Metchnikoff from what was real—or to guess what Flexner may really have thought. A somewhat more objective report comes from Dr. Kaiichiro Manabe, who later became head of the Institute of Physiotherapy at the University of Tokyo, and who was studying in Vienna at the time of this congress.

Manabe wrote that Japanese were very popular in Austria because of their victory in the Russo-Japanese War; Austrians "felt that they had been granted an extra ten years of peace" by Japan's defeat of Russia, and he said that Japanese were shaken by the hand in the streets of Vienna whenever they went out. He came upon Noguchi accidentally at the en-

---

*See Chapter 15, note 31.
**Italics mine.

trance to the university, and became Noguchi's guide. A large crowd attended Noguchi's first lecture, and people followed Noguchi out of the hall
with Manabe trying to clear the way. "There was a phone call from Dr.
Müller and I told Noguchi that Dr. Müller wanted to see him and he said,
'the great and famous Dr. Müller can't be wanting to come just to see me,'
so I checked again and told him. . . .When he heard this he started dancing around the room saying 'Müller of Munich is coming to see me!' "[23]
When Müller arrived, the crowd gave way to let him through and it was a
dramatic encounter. Müller and Noguchi greeted each other politely,
"both shed some tears," and both were very humble but delighted, according to Manabe's observation. Müller invited Noguchi to give a lecture in
Munich, Noguchi accepted, they both bowed several times, and everyone
was happy.

"At one of Noguchi's lectures," Manabe continues, "a young scholar
from the Koch Institute raised a question. Noguchi became very angry and
said 'How long have you been studying bacteriology?' and tore up the
young man's name card in front of the group. I was embarrassed, but
pleased at Noguchi's courage in tearing up someone's name card in
Europe."[24] The congress ended with a banquet attended by the prince in
place of the emperor, and Noguchi was even invited in a small group to
have lunch with the Emperor Franz Joseph.[25]

All the other Japanese at the meeting were amazed and some ventured to
ask Manabe who this person, Noguchi, was. As he was leaving Europe to
return to New York, Noguchi wrote a long letter to Chiwaki in which he
described the trip in detail and spoke of the position in which he now
found himself. Noguchi was euphoric and boasting as ever, yet now it was
all true. "I was quite popular in Vienna and probably because of this many
medical societies extended invitations to me. Those which I accepted were
Munich, Frankfurt, Copenhagen, Christiana,* Bergen, Stockholm, London, and Berlin," Noguchi wrote. He reviewed his scientific reports to the
congress, and he could report that he had received medals from King
Christian X of Denmark and the King of Spain, and a rumor of still
another from the King of Sweden. "This will be a great weapon for my life
in a foreign country, especially an anti-Japanese country like America, and
and will help me in competing with foreigners." Noguchi told Chiwaki that
he had visited ten cities, given eleven lectures, and had been the honored
guest at thirty-eight banquets. He had met members of the imperial
families of two countries, Austria and Sweden, and had been entertained
at dinner by the Japanese Ambassadors to Germany and Sweden, and the
Japanese Consul in Vienna.[26]

---

* Name changed to Oslo in 1925.

Hideyo Noguchi and Thorvald Madsen, 1904. Property of Dr. Sten Madsen

It would appear that the only players in this high drama who were not dazzled into incompetence were the young man from the Koch Institute whose name card was publicly shredded and possibly Professor Metchnikoff whose opinions were transmitted only through Noguchi. How could such a thing have happened?

Various people in later years accused Noguchi of simply lying about his findings, but this explanation is not sufficient. Much of the work he presented in Europe was sound, and no one realized yet that his culture of *T. pallidum* was questionable. Beyond that, Noguchi exposed his shortcomings in his own words. He drew his premature conclusions in reporting the transmission of syphilis to rabbits when he had been successful in only one out of thirty-six cases, promising follow-up work that never appeared. Similarly, in his guarded report of the rabies work, which was never amplified by either positive or negative data, Noguchi's optimistic interpretation of his observations soon became clear to everyone.

We do not know what he said in his verbal presentations in Europe. Although we have learned from Noguchi's own words that Theobald Smith was dubious about the Negri bodies in his rabies material, he rationalized Smith's doubts by saying that only the man who had worked with the material over a long period of time could make a definitive evaluation, and he further sought to beguile Flexner with, "I believe you are convinced. . . ." When Noguchi promised Flexner that he would pursue the investigations further after he returned from Europe, he did so with the qualification, "but you know that this is to convince others who are skeptical." Noguchi could not allow conservatism to prevail, at this crucial moment, over his desperate desire that things be as he wanted—needed—them to be.

Following the Vienna Congress and the European lecture tour, two letters were written that suggest a further reason why others allowed themselves to find Noguchi's apparent convictions so persuasive. One letter was addressed to Rockefeller by the secretary of the Royal Society of Medicine in London, and the other was addressed to Flexner by Professor Paul Ehrlich, director of his own institute in Frankfurt and winner of the Nobel Prize in 1908.[27] Both letters were lavish in their praise of Noguchi, and both writers took this opportunity to express their gratitude to Rockefeller for his financial support of science. On the strength of these letters, Jerome Greene once again took up his pen to call Rockefeller's attention to this evidence that his money was being well spent. "The Germans used to be incredulous about his [Noguchi's] work, but they now acknowledge its rigidly scientific character and have given him generous praise. . . .His success is so clearly due to the facilities provided and the wise supervision afforded by The Rockefeller Institute through Dr. Flexner that I am sure you and Mrs. Rockefeller cannot help feeling very deeply gratified."[28]

There is little question that Noguchi's career had reached its zenith even though his ambitions were never satisfied. On January 18, 1914, he wrote Kobayashi: "Since my return to the U.S., friends and others have congratulated me on my success in Europe and Dr. Flexner also is very pleased. . . .I hear that newspapers have talked of my getting the Nobel Prize this year, but this is just a rumor and no one knows. . . .My work may be chosen but there are many great scholars working to get this prize so a young person like me may not have a chance. My turn may come within five years."[29]

His triumphal reception in Europe enabled Noguchi to achieve his promotion to the highest rank at the institute. He was offered a position as director of a new laboratory that was being planned in New York, at a salary of $6,000 a year. He told Kobayashi: "I have talked with Dr. Flexner about this and he suggested that since I am one of the most important members of this Institute I should stay." He was made a full member at a salary of $5,000, and again he described in detail the pension benefits that went with this rank. Listing the names of six men who already were full members, he added his usual personal observations about their ages as well as his intention to start saving money and to help his family.[30]

Dr. Michael Heidelberger, who was a young man at the institute at this time, remembers Noguchi saying of his promotion, "Perhaps I shouldn't have done it. I may be too young."[31] Heidelberger was considerably younger than Noguchi, and he felt that this was an indication of Noguchi's true modesty. Heidelberger believes that Flexner could not have been forced to grant Noguchi full membership in the institute staff by means of an offer from the new laboratory; that Flexner would never promote a man who wasn't completely worthy.

To say this, however, is to shift the whole focus onto Flexner's judgment of worthiness and, concomitantly, to Noguchi's own judgment about his own work. If Noguchi's promotion could have been considered only on the basis of his achievements up to that time, which are recognizable by the strictest of standards as one reads his papers over sixty years later, and if one could estimate his potential value to the institute and to science in the light of these accomplishments, it would appear that he might well have been found "worthy." The snake venom work, the confirmation of the Schaudinn-Hoffmann discovery, the work in serum diagnosis (which was sound and helpful at the time even though it did not yield the practical result he hoped for)—all were solid achievements. Furthermore, Noguchi's cultivation efforts with *T. pallidum* had already yielded valuable techniques for the cultivation and study of other spirochetes even though the claim of pure culture of that organism was announced with unjustified assurance. The culture of the spirochetes of a form of relapsing fever had come from the use of those techniques. And the identification of *T. pallidum* with paresis and tabes was valid. In fact, Noguchi had accomplished, in less than fourteen years, an impressive amount of creditable work, flawed only by the excessive optimism of his own interpretations.

But this was not all he did. There were also the ill-considered claims of more sensational accomplishments, given credence by the still unresolved problem of the pure culture of *T. pallidum*. Flexner and Smith had both been doubtful of his work on rabies, and Noguchi knew that they were. There was also his work on poliomyelitis: a small, but critical, addition to Flexner's project: one on which Flexner worked, and that he directed and defended long after Noguchi's death and which was discredited in the end. But these alleged triumphs by Noguchi were in fields with broad public appeal, and they contributed more heavily at the time to the evaluation of his worthiness than his truly substantial accomplishments did.

It must be assumed that Noguchi's qualifications for full membership on the staff of the institute included his ability to draw acclaim for the institute and goodwill for Rockefeller. If he hid his own ambition behind his humble and modest face, he could surely see that his technique was effective. He had learned very early that, in Japanese society, one is judged and privileged by one's personal connections. Following the same guiding principle in the West, Noguchi offered his willingness to tread on dangerous ground, he put a humble face on it and let his connections work for him. He questioned only whether he had moved too far too fast.

When Noguchi was in Vienna, an artist, Emma Löwenstamm, made an etching that portrays Noguchi as no photograph has ever done. His face, shown in profile, is that of an innocent child; yet, seemingly imposed and out of place on that baby face, are full greedy lips whose sensuality is only partly concealed by the tight little mustache above them. His colleagues described him as childlike; Clark refers to "our little mop-haired Japanese scientist."[32] No one mentions his greed, much less any perception of the deprivation that might have formed that greed, and no one would see his sensuality. But several colleagues at the institute acquired prints of that etching. Landsteiner once brought one from Europe as a gift to Peyton Rous, who kept it hanging on the wall of his office until his own death some fifty years later.[33]

Flexner was wont to confide in Rous about Noguchi, and we can believe that Rous knew Noguchi as well as anyone at the institute did. Rous refused to discuss Noguchi, saying he knew nothing, but he may have known one of Noguchi's secrets that was revealed to Flexner for the first time in 1913. Noguchi learned himself for the first time that he had a heart enlargement which, on further examination, was found to be aortic insufficiency growing out of an old and inadequately treated syphilitic infection. In 1914, when he was first examined by the institute cardiologist, Noguchi had a positive blood Wassermann reaction.[34] One story has it that

Hideyo Noguchi. From an etching by Emma Löwenstamm, Vienna, 1913. Property of the author.

he learned of the enlargement when he and some others were trying out an electrocardiograph apparatus that had been built for experimental purposes;[35] such machines were only in the process of development at that time. Another story, one that Noguchi preferred, was that his heart condition was discovered in the course of a medical examination for life insurance.[36]

On January 19, 1914, Noguchi wrote to Flexner protesting a suggestion that he be treated by physicians at the institute hospital.[37] He preferred Drs. Libman* and Longcope of Mount Sinai Hospital and wished to make his own choice in the matter. Flexner noted on the margin of this letter

*Dr. Emmanuel Libman served as Noguchi's personal physician from this time forward.

that he had agreed to Noguchi's choice; but since Noguchi was regularly examined thereafter by Dr. Alfred Cohn, the cardiologist at the institute, he may have done this as well to please Flexner. The institute was a gossipy place, as most such institutions are; yet, although everyone knew that Noguchi had a heart problem, the secret of its origins seems to have been kept.

Noguchi was essentially asymptomatic and remained so for the rest of his life. According to Dr. Cohn's records, later blood examinations never again turned up a positive Wassermann reaction.[38] We cannot tell whether the administration of Salvarsan by Dr. Longcope or the prolonged fever during Noguchi's siege of typhoid fever in 1917 should be credited for arresting the progress of the tertiary syphilis. The physicians who treated Noguchi during the typhoid illness were concerned about his heart enlargement, and Flexner was reluctant to have him push himself on various expeditions for the same reason, but it never seems to have created a problem.

We have no idea when Noguchi contracted the syphilis or what treatment he had received during the primary stage. It seems very unlikely that he had not known, even before the discovery of his heart enlargement, that his blood Wassermann reaction was strongly positive. He would surely have tested his own blood in the course of his experiments. Okumura wrote that Noguchi used his own blood in some of his work, which now appears unlikely, but we can see why he needed so much blood from his compliant young friend Araki.[39] But beyond such details, as Noguchi worked with such complications of syphilis as tabes and paresis, we marvel at the possibility that he had done nothing about his own condition. Did he consult none of his physician friends? Was his embarrassment so great that he preferred to gamble on the statistics? "About two-thirds of the total untreated infected individuals go through life with minimal or no physical inconvenience, although more than one-half of these will remain serologically positive for life. . . .Unfortunately there is no known means of predicting which patients will develop the late lesions of syphilis and which will not."[40]

Noguchi called himself a *kiwadoki yatsu* — a man who lived by taking chances. It would have been entirely in character with the rest of his behavior that he took this chance rather than risk the exposure of this souvenir of his less decorous days.

14

Return to the Orient

The fact is that heathen nations are being everywhere honeycombed with light and with civilization, with modern industrial life and application of modern science, through the direct or indirect agencies of the missionaries. . . . The result will be eventually to multiply the productive power of foreign countries many times. This will enrich them as buyers of American products and enrich us as importers of their products. We are only in the very dawn of commerce, and we owe that dawn, with all its promise, to the channels opened up by Christian missionaries.

—Reverend Frederick T. Gates[1]

IN 1915, NOGUCHI AND FLEXNER BOTH VISITED JAPAN—BUT SEPARATELY. The timing of their visits was close, and they both brought offers of increased contact with American medicine; their offers were slightly different in the sponsorship under which they came, however, and in the rewards they sought for their generosity.

Noguchi came under the sponsorship of personal friends in Japan, and thus his visit was personal as well as professional. He offered his scientific reputation, experience, and talent to his country for which he sought recognition in the only form that meant anything: an academic position. Flexner came as a Rockefeller Foundation man, en route to China for the dedication of the Peking Medical College, which had just come under the wing of Rockefeller philanthropy. He stopped in Japan to offer the benevolence of his sponsors to Japanese scientists for which Flexner, too,

149

sought recognition: not for himself, but for the purity of the stated objectives of the Rockefeller Foundation.

Noguchi returned from his visit embittered by failure, the extent of
which he was only partially aware of when he left Japan. Flexner returned in
triumph, which had been gained, in some measure, at Noguchi's expense.

Noguchi's Trip. When the war broke out in Europe in 1914, the whole
picture of scientific interchange was altered; for Noguchi, the effect was
even more immediate and profound. On the European side, all of his new
contacts with scientists and with famous laboratories vanished; he even felt
a difference within the membership of the institute in New York. As he
wrote to Chiwaki: "Even though German scholars, British people, and
Russians are eating together at this institute, they don't seem completely
comfortable with each other. As for the academic world, it has become
deserted without the European scholars. It is as if the world had suddenly
shrunk."[2] But the implications for Noguchi of Japan's involvement in the
war were greater yet.

In the 1890s, Japan had begun to gain a foothold on the Asian continent, but when the so-called Triple Intervention of Germany, Russia, and
France had forced Japan to relinquish much of what she had gained, Japan
considered that she had suffered a national humiliation. In the following
years the maneuvers of all these powers to further their interests in China
and in the Far East led to an Anglo-Japanese Alliance in 1902, and to the
Russo-Japanese War in 1904. Although Japan's interests in the Far East
were acknowledged in the treaty ending the Russo-Japanese War in 1905,
and the Anglo-Japanese Alliance was renewed and extended in that same
year, Japan's influence in China was still restricted by Western powers.
When World War I broke out, Japan immediately came forward to honor
her commitment to Britain by declaring war on Germany. She moved
quickly into China, achieved a rapid victory, and when three hundred German prisoners were brought to Japan there was national rejoicing. The
long-standing humiliation had been wiped out; Germany, once the
respected and feared symbol of the power of advanced nations, had been
defeated by Asians. Privately, however, it was seen that there was a price;
Germany now expelled all Japanese students from the Austro-Hungarian
Empire where there had been fifty Japanese students in the field of
medicine alone.

Baron Ishiguro, the elite patron of Western medicine in Japan, had
foreseen this possibility; detached from the factionalism of academia, he
not only kept himself informed, but exercised the power of decision in the
background. When Noguchi had first acted as Flexner's interpreter in
1899, Ishiguro had sent for Noguchi to learn of Flexner's impressions of
Japanese medicine. He had kept in touch with Noguchi's progress, had
written congratulations on his success in Philadelphia, and had read all the

reprints that Noguchi never forgot to send him. Now Ishiguro thought of America and of Noguchi, and he suggested that, since Noguchi had been so highly regarded on his European tour, it might be appropriate for the Japanese also to make some formal acknowledgment of his achievements.[3]

This suggestion was not greeted with unanimous enthusiasm among members of the various factions. They were already having trouble in negotiating solutions to internal problems, and since none of these cliques had ever shown any enthusiasm for Noguchi, it was not easy for them to agree on a plan for increasing contact with American medicine by embracing Noguchi. But no face-saving compromise was available. Noguchi's acceptance in Germany and Baron Ishiguro's prestige left no alternative. Thus we find Noguchi writing to Kobayashi:

> Dr. Chiwaki informed me that my application for the degree of D. Sc. was approved unanimously by the professors of the University of Tokyo. My dissertation was submitted on May 22 and was approved on June 29 which is a very quick decision compared with last year's medical degree. There is such a difference — like clouds and mud. It took me two years to get the medical degree. . . . .In February of this year Baron Kikuchi* wrote me the secret notice that I might be receiving the Academy Prize (very secret). At that time I refused with regret that the opportunity is not ripe yet and I did not want to get in the way of scholars in Japan whom this prize is meant to encourage. However, I would be very happy to accept it if I am chosen unanimously.

Noguchi understood the dispute and for this reason he held out for an award of the Imperial Prize by a unanimous vote of the Academy; he nudged a little by telling Chiwaki that there was a rumor that he had been nominated for the Nobel Prize. He was planning his strategy with great care.

By December 1914, even though Noguchi admitted to Chiwaki that he was not going to get the Nobel Prize that year, he set his views forth in such a way that they could be passed on to those who might be talking about him. He acknowledged the rights of Japan in Asia and expressed his sympathy with Japanese objectives: "the first step must be to drive European and American forces out of the Far East." He spoke of the need for development of Japanese military and economic power sufficient to "surpass the European power for invasion," and suggested the need for Japan to start training her own students rather than spending so much money sending them abroad. He mentioned the transfer of the Kitasato Institute from one ministry to another, and though he did not know whether such a change would make any real difference, he felt that the crucial test of such matters was to be found in the quality of the training programs.

Noguchi also spoke of the language problem and the need for the

*President of the Imperial Academy.

Japanese to acquire proficiency in a Western language in order to get their work more widely publicized. "No matter what one achieves, it is of no use without world recognition. No matter what language we use for publication, it need not affect our national pride." He expanded on his views of Japanese shortsightedness in the past and the wisdom of yielding a little to win a lot—a popular idea in Japan.[5]

Noguchi's friends were now mobilized to contribute to his strategy. Hajime Hoshi, who had started sending monthly payments to Kobayashi for the purchase of land for Oshika, now promised 5,000 yen to pay the expenses of a trip to Japan for Noguchi. In view of the expected Imperial Prize, Hoshi thought it did not sound right to say that Noguchi was coming at his invitation, "so we made it look as though he was coming to see his mother."[6] Ishizuka, too, was called into service. He was now a dentist, practicing and teaching at Chiwaki's dental school, but with offices at Nagaoka and Niigata as well. An accomplished amateur photographer, Ishizuka went to Inawashiro and took pictures of Oshika that were sent to Noguchi to show how much his mother had aged. Ishizuka wrote that if Noguchi expected to see her again, he had better come soon.[7]

When Noguchi wrote to Chiwaki in May 1915, he thought that he had outwitted the academicians:

> A short time ago I received notice from the Imperial Academy that I am to receive the Imperial Prize which surprised me. I knew that my work had achieved some recognition in the academic world but I doubt that it is worthy of the Imperial Prize. . . .This is a great honor for me. I would like very much for you to be present at the ceremony in place of me. Please show the plaque to Dr. Watanabe, then send it to my foolish mother through Sakae Kobayashi. . . . Then, when you have time, please send it to me. . . . I would like to go home to Japan now, but Japan is so far away and my present circumstances do not allow me to do so. But I will try to find a way somehow to return to Japan within a few years. I confess this in order to let you know how happy I am today.[8]

As late as July 4 Noguchi indicated to Kobayashi that he would like to visit Japan in the fall, and said that Ishizuka had not sent the pictures of Jōbo yet.[9] Was the plan still not complete? Was Noguchi still not sure that Hoshi was going to pay for the trip? But on July 22, Kobayashi wrote that the plaque announcing the Imperial Prize had arrived and a formal celebration was being organized in the village to put it on display. Hoshi had written Kobayashi that Noguchi's visit was now a certainty, and Kobayashi plaintively added that Noguchi could now have any girl in Japan he wanted; if he would only tell them, they would surely find her.[10] On July 21, barely three weeks before his departure, Noguchi finally wrote that he would arrive in Japan on September 5.[11] Hori, his friend in New York, told later that when the money from Hoshi arrived, Noguchi put it

Oshika Noguchi, 1915. Hideyo Noguchi Memorial Association

in his pocket and started for the ticket office to pay for his passage, but when he reached for the money it was gone. He retraced his steps, and found it on the street where he had dropped it.[12]

Clothes-conscious as he always was, Noguchi set off with a new formal morning coat, striped trousers, and a top hat, as well as less formal attire, and he carried the necessary gifts. We are told repeatedly that Noguchi's trip had no connection with Flexner's trip, but since the Flexner party included Welch and other important people from the Rockefeller Foundation, it can be assumed that Noguchi wished to be seen with them in Tokyo and had been trying to arrange his trip at a time when this would be possible. He wrote Kobayashi from shipboard a few hours before docking:

> I am due at Yokohama at 5 p.m. today. My heart is filled with ten thousand feelings at this moment. . . .I am planning to stay temporarily at Imperial Hotel with Hoshi's help but I may change and go to Dr. Chiwaki's if he wants me to do so. I think it is necessary to stay at a hotel since I have some business with foreigners. My return will be determined by Dr. Flexner's stay—he is visiting Japan from the U.S. and I have to stay in Tokyo and take care of his party until they leave for China.[13]

Newspaper reporters swarmed onto the ship, followed by his friends. After the greetings and interviews were over Noguchi got into a two-horse carriage, which had been supplied by Dr. Rokkaku of Inawashiro, and was driven to the Yokohama Railway Station. The bilingual newspaper *Yokohama Asahi* quoted Noguchi as telling reporters: "Among other adventures I have lost some fingers. While engaged in research work on snake venom in the Pennsylvania University, I was bitten by a very poisonous snake and there was no remedy but prompt amputation." He spoke about Japan's being cut off from Germany by the war and said: "America already claims superiority over Germany in regard to hospital organization and equipment, surgery, and bacteriology, and is probably ahead in other respects also."[14] Of the ten thousand feelings that filled his heart, these are the two that came out.

Everyone who knew Noguchi before he had left Japan had surely been aware of his deformity, and probably most of them knew how it had happened. For those seeing his deformed hand for the first time, Noguchi chose the role of the adventurer, accustomed to taking snake bites and amputations in stride as mundane events in a life of heroic exploration, to help him in controlling the complicated emotions that would be unavoidable in the days to come. It recalls the story related by Grunberg—of Noguchi's having told that Oshika had burned the hand as punishment for his neglect of his studies. Noguchi very likely told Grunberg that story, and may well have thought of it as he approached his reunion with the mother who had somehow given rise to it.

At Tokyo Station, Noguchi faced a mob of greeters. The station itself was still a symbol to Japanese; it had been completed less than a year before, and its dedication had been combined with the recurrent celebrations of the victory over the Germans in China. As Noguchi was met by important officials, his arrival took on the semblance of another such celebration of the success of a Japanese in a Western accomplishment. The timing could not have been better and the newspapers made the most of it.

Every prominent Japanese academician was there but one — Dean Aoyama of Tokyo University, who still recognized only German science, and beyond that, regarded Noguchi as a Kitasato man and not even a particularly good one. Noguchi did not recognize the slight since Manabe, from Tokyo University, was there. Manabe, the man who had become Noguchi's guide in Vienna in 1913, had come to welcome Noguchi as a personal friend rather than as a representative of the university. Now, once again, Manabe's tact and resourcefulness were of service to Noguchi. Realizing at once that Noguchi thought he represented the university at the station, Manabe scurried to Dean Aoyama's office and persuaded this haughty man to change his mind. He told how Professor Müller, "the Aoyama of Germany," had warmly recognized Noguchi and his achievements, and Aoyama betook himself to the Imperial Hotel and paid his call.[15] Baron Ishiguro emerged from his power base behind the scenes to pay a call, and the next day Noguchi went the rounds and returned the calls. Protocol was observed at the highest levels, but Noguchi must have been shaken when he learned that Flexner and his party had already come and gone.

Okumura writes much more of Noguchi's reception by the general public, by his home town, and by his family, than of his reception by the academicians. Noguchi may have suspected, before he left Tokyo two days later to visit his family, that the formalities that had been observed on his arrival were almost the only efforts that the university men would make on his behalf. On his second evening, however, he and Kobayashi joined a reunion of personal friends at Chiwaki's house and plans were made for his trip to Inawashiro.

Kobayashi had brought a detailed set of instructions that he said came from Oshika, and Noguchi followed them to the letter. He was met in Koriyama by a delegation that rode the last part of the train journey with him. The festivities had started in Inawashiro some three days before his arrival, with fireworks and speeches every night. Noguchi wore his formal morning coat with the gray striped trousers and a straw hat, and in this costume he was drawn through the village in *jinrikisha*. A prescribed route was followed so that everyone could see him, but his family remained behind closed shutters in their house. Oshika thought that they should refrain from showing excitement in front of the villagers. Noguchi toured by the Hachiman Shrine where he gave thanks for his protection and he spoke there to the people, giving his humble greetings. He visited each of the thirty houses of Sanjōgata Village, thanking these neighbors for helping his family in his absence, and he went to the cemetery and reported his return to his ancestors. Last of all, he went to his own home where his whole family—including Sayosuke who had come from Hokkaidō for the occasion—had now come to stand outside their doorway to greet him.

Noguchi talked with his mother long into the night, and devoted the next few days to local banquets, speeches, and visits. He went over the property and financial transactions with Kobayashi and, however they were settled, all parties were well satisfied; the family was now considered well off, secure, and out of debt.

One evening, when Oshika, Noguchi, and Mr. and Mrs. Kobayashi were alone at dinner at Kobayashi's house, Noguchi brought out a large picture of his wife and showed it to them. It is almost impossible to believe that this is the first they knew of his marriage and that his father was excluded from this gathering, yet that is what we are told. They discussed his marriage at length; Noguchi told them that Mary was gentle, like a Japanese woman, and he thought that they would like her. They talked about her height and Oshika remarked that Mary could carry him on her back like a baby since he was so small.[16] There is hardly anything more Oshika could have said. Her son had been married for more than two years, he had come home with undreamed of glory for the family, he had—directly or indirectly—established their financial security, and he certainly was not going to return to a permanent life in Aizu. Oshika was in no position to complain, no matter what she may have thought of his having married without consulting the family, and no matter what the implications might be for the future of the family.

There was a special dinner party at which the friends who had known Noguchi since his childhood joined together to form the *Chikubakai,* a club whose whole membership dedicated themselves to the preservation of the story of his origins. Kobayashi was undoubtedly one of the moving spirits of this club, which eventually made something of a local industry out of the collection and preservation of Noguchi memorabilia. *Chikubakai* played a later role in the establishment of the Noguchi Memorial Foundation and probably formed the nucleus of the group that approached Ide, in 1955, to write the biography which was never written.

We may recall Ide's impressions when he visited Inawashiro and saw "the awful house in which the Noguchi family had lived."[17] For most of the nights of his visit, Noguchi shed his fancy Western clothes and slept in this house, on the dirt floor with the pill bugs, inhaling the stench. He made a special call to bring gifts to the family of his benefactor, Yago, whose son told of it to Ide. The son had been a child at the time, but the incident was clear in his memory. The children were lined up to greet Noguchi and were impressed by the courtesy with which he returned their greetings. Then their grandmother was brought in, she being the one who had been known to Noguchi as the nagging mother of his boyhood friend, Yasuhei. Now Noguchi made a speech about the great amount he owed to this family, and took out the gifts he had brought them. There was a gold watch and a box of Sunkist oranges. The old grandmother greeted these with noisy contempt and threw the oranges all over the tatami, while Noguchi sat with bowed head listening to her tirade. The old lady was dragged out amid the embarrassment of the other members of the family, and her grandson remembered, some forty years later as he told the story to Ide, hearing her shrieking, "And the Inbahnesu! What happened to the Inbahnesu!" Her grandson did not know at the time about the pawning of the Inverness coat that had been made for Yasuhei's wedding; but Noguchi remembered all too well.[18]

He went for a few days with Kobayashi to an Aizu hot spring where they talked at length about matters that bothered the old country samurai. He worried about Noguchi's working to find a cure for syphilis; he took the attitude that syphilis is a punishment that is visited on people because of their immorality and asked why Noguchi wished to find a cure, when curing the disease would have the effect of increasing immoral behavior.[19]

With such thoughts in his mind, with the picture of himself as the hero who had restored his family to the level at which he now saw them, with thoughts of the mother whom he had once accused of burning his fingers off to get him out of this environment and whom he now saw surviving in pride and gratitude although almost dead from the efforts she had made, and with the accusing shrieks of the old Yago grandmother ringing in his ears—Noguchi left for a series of banquets and lectures in Niigata on the other side of the mountains.

Ishizuka accompanied him on this part of the trip. After it was over he received a letter that Noguchi had written on the train and left for delivery after his departure. More saddened than satisfied by the visit to his home, Noguchi nevertheless thanked Ishizuka for having taken the pictures of his mother and spoke of their influence in bringing him back to Japan for which he was grateful. He enclosed another of his sad, Chinese-style poems:

The fire-wheel clanking on while
intercepting the long wind
Bits of the mountain and part of the river
are now in view.
The cloud, chilling, and the pines, elegant.
In this area of the red bridge,
We are riding above the azure sky.[20]

Back in Tokyo, Noguchi spent an evening with Miyabara, who found him so melancholy that he suggested their going out on the town. Noguchi openly wept as he said that he would never see his mother again after this trip, and perhaps would never see Japan or any of his friends again. Walking down the street, they passed a *Naniwabushi*, a "musical recital of ancient tales" in the style of the old minstrels, a beloved diversion of country people and one that Noguchi had known and loved. He wanted to go in and they did; but he was recognized by the audience and his presence stopped the show so they had to leave. More and more he was finding himself out of place, and no academic overtures had yet been made to him.

Noguchi wrote to Kobayashi and asked him to bring Oshika and Mrs. Kobayashi to Tokyo, and invited them to accompany him on a ten-day tour of the Kansai area where he was to make a number of appearances. He has been accused of showmanship in taking his old country mother on such a tour. She had never been out of the Aizu area, and Japanese men did not take their wives, much less their backcountry mothers, on such tours. But showmanship or not, Noguchi did it, and Oshika enjoyed every minute. The things she saw, the food she ate, were new and strange to her. He took her to a banquet at which one of the most famous *geisha* of Japan entertained, and the *geisha* fled the room in tears, overcome by the sight of his gentle courtesy to this worn-out old peasant woman.

Noguchi had given his father money to return to Hokkaidō and had spoken with Dr. Rokkaku about his father's health. Noguchi told the doctor that Sayosuke was so much worse off when he could not drink that it was of no benefit to try to stop him, and he urged Rokkaku to step in only as complications arose. Inu and Sayosuke had stayed in the background together during all the celebrations as they felt their presence might be a source of embarrassment.

Noguchi's first love, Yoneko Yamanouchi, attended the celebrations in the Aizu area and accompanied him on some of his local excursions. She now had two children, her husband had died, and she was trying to practice medicine and bring the children up alone. She asked Noguchi to take her two sons to the United States and have them educated there, but he explained that it was too difficult for foreign students to maintain themselves and he could not afford to support them. There are differing stories of Noguchi's expectation of seeing Yoneko still another time before he left and his disappointment that a further meeting did not come about. Perhaps none of the stories is true, but each adds to the growing feeling that this trip was entirely one of farewell.

On the academic level, the only banquet in Noguchi's honor was given in Tokyo at the Kitasato Institute, a nominal observance of his presence in Japan. With this one exception, the invitations to banquets and to lecture came from local medical societies, business groups, and students; none came from universities.[21] In his lectures, Noguchi talked about American customs and medical practice. He told of the support of research by wealthy people in America, and he pointed out differences in both security and freedom between Japanese and American scientists. Although he privately deplored the fact that Japanese scientists turned to politics once they became successful and said that their scientific work deteriorated as a result, Noguchi made no critical remarks in public. He told newsmen that everything in Japan seemed to be going well. He observed that scholars had some difficulty in getting their work published in the West and thought he could help with this problem on his return.

On September 21, Noguchi wrote from Tokyo to Flexner in China: "As I told you, I am thinking to make a few trips to certain localities where there are some obscure endemic diseases (such as *tsutsugamushi* and Weil's disease). I am told that they had just found something which they believe to be the etiological factor in these diseases. Yet I have to see myself before accepting."[22] Noguchi added that he planned to meet the Flexner party on their way back from China and to conduct them on sightseeing around Tokyo; he even hoped that he might return to the United States on the

Hideyo Noguchi, Morinosuke Chiwaki, Mrs. Chiwaki, and the Chiwaki children, 1915. Hideyo Noguchi Memorial Association

same ship as they would take. His cool reception by the Japanese academicians made him all the more anxious to be seen with the prestigious Rockefeller group with which he was known to be associated.

Meanwhile, Noguchi continued his tours and lectures and visited the laboratories around Tokyo. He learned that a spirochete had been identified as the causative agent of Weil's disease by a group of workers at the Imperial University on the island of Kyūshū, some seven hundred miles southwest of Tokyo, and that the work was being extended at the Kitasato Institute and in the laboratories of a hospital in Chiba, the prefecture adjoining Tokyo. When Weil's disease (infectious jaundice)* was first described in Germany in 1886, its characteristic symptoms were said to be fever, jaundice, and hemorrhage. By 1915, it was recognized as a prevalent disease among certain occupational groups in Japan, and among soldiers in the trenches in Europe. Noguchi not only saw the evidence that the organism had been correctly identified; at Chiba, he also made suggestions that led to the pure cultivation of the organism. His assistance was acknowledged by the Chiba investigators when they published their work.[23]

He also visited the old Kitasato Institute, now a part of the University of Tokyo, and saw the work in progress on rat-bite fever and on *tsutsugamushi*. Rat-bite fever is just what the name implies: an infectious disease usually contracted from the bite of a rat. Miyake had introduced the name (*sodoku*— "rat poison") into Western usage through his German publication in 1900, but the disease is known throughout the world. It is caused by a spirochete that was identified by K. Futaki at the University of Tokyo, probably before Noguchi's visit. *Tsutsugamushi* literaly means "insect disease," an indication that insects were already recognized as the carrier. The disease is common in the Orient, but not in America and Europe. In 1915, its causative organism was still being sought and there was hope that, when found, the disease would respond to the cultural techniques that were proving useful with spirochetes.

Noguchi was forced to acknowledge the high quality of all this work, but he must also have learned that Flexner had already seen some of it and had already accepted some papers for publication in the *Journal of Experimental Medicine* of which Flexner was editor. Perhaps Noguchi did not believe this at first, and only gradually came to realize that it was true. The only clear-cut offer he could make to the academicians, to use his influence on their behalf, was now wiped out, and he suddenly decided to leave before Flexner and his party returned to Tokyo.

Noguchi spent his last day in Japan making farewell calls, and he visited the grave of Dr. Kawakami, whose acquaintance with Kitasato had been of

---

*Infectious jaundice (Weil's disease) is caused by a spirochete, and is not the same as infectious hepatitis, which is caused by a virus.

such help in gaining his admission to that institute over fifteen years before. In the evening he joined with friends in a farewell dinner, again expressing the belief that he would never see Japan, his family, or his friends again. But certain as it seems that his final rejection by the academicians had reinforced these conclusions, we find no record of anything he ever said about it.* It would seem that this lack of acceptance was all the more insulting to Noguchi in the context of the great public fuss that had been made over him.

On the morning of November 4 he wrote a letter explaining his early departure to Flexner: "It was a whirlwind trip with no real rest at all. I was carried from one place to another and could not do as I originally intended to. There were nearly sixty dinner or luncheon parties within the past two months and I cannot stand any more. I have also had occasion to study Weil's disease and *tsutsugamushi* disease and will tell you something about them when we meet together at the Institute."[24] He sailed that afternoon to travel back alone.

Flexner's Trip. Dr. Flexner and Dr. Welch had come to the Orient to attend the dedication of Peking Medical College under its new auspices, the China Medical Board, an event that John D. Rockefeller, Jr., considered of such great importance that he joined the group himself in China in time for the ceremony. Rockefeller's father, who had built the family fortune, already had a long history of support of Baptist foreign missions and of educational institutions within the United States. The Reverend Frederick T. Gates had been placed in charge of the Rockefeller philanthropic activities when they became too large for the senior Rockefeller to manage them to his satisfaction. The establishment of the Rockefeller Institute had been only the first of Gates's creative designs. When the son took over the major responsibility for the Rockefeller enterprises on his father's retirement, he pursued the philanthropic affairs of the family with special devotion, but still in collaboration with the Reverend Gates.[25], [26]

The Rockefeller Foundation, which was established in 1913, was more a creation of the son than of the father; and the China Medical Board was an outgrowth of the foundation. Although the Standard Oil Company had led the field in marketing fuel oil in the Orient in the 1880s, and the *Meifu* (American fuel) lamps that the company gave its dealers to distribute in China were known throughout the country,[27] the particular interest in medical education in China at this time was related to the establishment (1911) of the first republic in the long history of that country. Convinced that the commercial and industrial policies of modern China would be determined by the first generation of students to receive a modern education under the new republic, the Rockefeller Foundation viewed educational and missionary work in China as an "investment in leadership."[28]

Noguchi's letters to Chiwaki written after this trip are said to have been destroyed in the 1923 earthquake.

John R. Mott, who became the first chairman of the trustees of Peking Medical College, had been an aggressive advocate of American influence in education in China: "If we wait until China becomes stable we will lose the greatest opportunity that we shall ever have of dealing with the nation."[29]

Dr. Welch and Dr. Flexner returned to Japan late in November 1915 having spent three months in China. The prospects for the college and for achieving the goals of the foundation had appeared bright, and even while Flexner was still in Peking he had received a grateful response to overtures he had made in Japan.

On September 8, six days after Flexner had left Tokyo for China, Dr. Miyajima addressed a letter to him at Peking in which he thanked Flexner for his consent to publish a paper by Dr. Koga from the Kitasato Institute and asked if he would also arrange to publish a paper by Dr. Inada of Kyūshū University on the discovery of the causative agent of Weil's disease.[30] Flexner's reply, which was not written until after his return to New York, demonstrates the level of cooperation he had offered.

> On the way across the Pacific I was able to go over with care the first of Dr. Koga's papers, and I made corrections in the manuscript as I felt sure you would approve. . .In the meantime I am proposing to publish the paper of Inada and his associates without any delay. I found that Uhlenhuth has published in the *Medizinische Klinik* for October 31, 1915. He succeeded in communicating. . . so-called infectious jaundice to guinea pigs. He also found the spirochete. However, he seems to think the virus is filterable and makes no reference at all to Inada's work. I am therefore hurrying Inada's paper through, so that it may appear immediately, because our journal reaches Germany promptly. From that paper he will not only see that Inada has worked out the subject, but also how much he has covered. In this way priority will be established. I trust you will approve of this plan, which means that you will not receive the galley proof of the article, as that would involve great delay.[31]

Flexner recalled in writing of this trip in his diary how the Japanese had welcomed Noguchi, and he mentioned that Miyajima (of the Kitasato Institute) had told him, "if N had stayed in Japan, no matter what he had done, he would not have been so honored." Flexner said that this simple remark conveyed a great deal to him—that the Japanese had failed to appreciate Noguchi and now were jealous of him, to which he added, "Why not?" But Flexner also said that he had seen Noguchi over there, which seems to have been a false recollection on his part, and that he never expected that Noguchi would be "called back."[32] He had reason to feel some guilt at not having seen Noguchi in Japan, and to know that there was more than jealousy behind the Japanese not calling him back to an academic position.

Flexner had wound up his return visit to Japan with a major address at Kyoto Imperial University, whose professors had ignored Noguchi's visit

Japan. Prefacing his lecture with a review of his association with Noguchi he added: "It cannot help, I think, being a source of gratification to you all that so distinguished an investigator is of your nationality and received his training in medicine in your own country." Flexner then spoke of recent advances in medicine, and of the current understanding of viruses that he described as ultramicroscopic parasites whose disease-producing potential had been demonstrated by deductive processes. "In other words, the modern study of pathology of the infectious diseases has yielded to biology, as it were, a new class of living things. I have introduced this subject especially for the reason that only recently in conjunction with Dr. Noguchi it has been possible for us to show that one of these ultramicroscopic parasites can be cultivated definitely outside the body, and when it is cultivated it is no longer microscopic."[33]

He set forth his own erroneous conception of the "globoid bodies" of poliomyelitis, saying that Noguchi, using the technique which he had developed for the cultivation of spirochetes, had cultivated the poliomyelitis organism, and thus made it visible because, when cultured, it formed clumps. Flexner gave no indication that either he or Noguchi recognized any significance in the presence of living tissue that Noguchi added to his culture media as an essential part of his "special methods." As had already been suggested by their previous reports, both Flexner and Noguchi were so preoccupied with their peripheral objectives that the significance of the findings before their eyes quite passed them by.[34] Noguchi had his own problems, and Flexner looked at everything from the standpoint of its immediate value to the Rockefeller Institute and the Foundation. The work that Flexner had seen in Japan had nothing to do with viruses, but it was of interest because of the problems with rat-bite fever and Weil's disease in the trenches in Europe, and because of its potential for establishing confidence and goodwill among the Japanese scientists who were interested in new contacts with the West.

The *Journal for Experimental Medicine* had never published a paper from Japan before this time, but in the years 1916-18, eighteen papers from various Japanese laboratories appeared. Six of these papers were published with the notation that they had been "received for publication on December 1, 1915," when Flexner was still in Japan, a policy that was contrary to the stated policy of the journal.[35]

It appears that Baron Ishiguro's suggestion that Noguchi be given some honors as an opening wedge to American facilities had not been necessary after all. What Noguchi thought, whether he recognized the extent to which he had been caught in the middle and used by both sides, is a matter only for speculation in the light of the profound sorrow with which he left his country for the last time.

15

Starting Over

ONCE AGAIN NOGUCHI ARRIVED ON AMERICAN SHORES BEARING GIFTS, but this time there was no wreath of smiles. Among the four cases of souvenirs he had had shipped from Japan, one contained a personal gift from Kobayashi, a suit of samurai military dress of the Kobayashi family. Valuable in 1915 only for its historical, sentimental, and decorative interest, even a caricature of the society in which it had been taken seriously, the samurai garb still had the power to jeer at one whose family had never been privileged to wear it, to remind an ambitious peasant that he was a homeless upstart.

Yet Noguchi had learned much on this trip. He had learned that the Japanese bacteriologists had great competence and they had even invaded his own specialized field of spirochetes. Noguchi knew that the Japanese bacteriologists respected his work as they acknowledged the use of his methods in their publications.[1] He had learned that scientific interchange between the United States and Japan seemed desirable to Flexner and Welch as well as to the Japanese, but that they did not need him for this purpose.

Noguchi had a reputation to uphold, symbolized by his position at the institute, his publications, and his collection of "harmless medals," so he followed his old pattern of submergence in work when bitter feelings threatened to take over. An invitation to deliver one of the Harvey Lectures at the New York Academy of Medicine, an honor extended only to men who had made distinguished contributions to medicine, provided the impetus for getting Noguchi back to work. His subject was the spirochetes,

164

a subject that he could discuss as an acknowledged expert. He worked on the preparation of the lecture throughout the months of December 1915 and January 1916. He began with a detailed survey of the literature and continued with a presentation of his own work and the work he had recently seen in Japan. A comprehensive survey of the physical, chemical, and biological properties of these organisms, the lecture was so long and detailed (and his English still so difficult to understand) that many in his audience fell asleep. The lecture was published in full, however, and remained the definitive work in the field until it was replaced by Noguchi's own revision of it in 1928.[2]

Noguchi began to acquire students from Japan. The first was a cousin of Hoshi, young Akatsu, who had been forced to leave Germany; then Yamakawa, who came directly from Japan, followed by Ohira. Many of these visitors showed up without having made any arrangements in advance, merely bringing a letter of introduction as Noguchi had done. But they had support from home, and they had a Japanese man of status to turn to, a man who knew what was expected.

Ohira recognized that Noguchi had used his good offices with Flexner to obtain permission for these first Japanese students to work at the institute, but as others appeared during the next few years — Kudo, Matsumoto, Takano — word spread among the Japanese in New York that Noguchi treated them all badly. Flexner noted that Noguchi gave them "unimportant problems" and wrote that he finally discouraged Noguchi's taking Japanese workers.[3] Noguchi did treat some of the Japanese students very badly, but the minor problems with spirochetes that he assigned to others were educational for them and, in several instances, led to publication.

Akatsu, who had intended to stay at the institute only eight months, remained for almost two years. He found Noguchi irritable in the laboratory, unhappy, and admittedly searching for something big but not finding it.[4] Ohira and Yamakawa took turns getting very angry with Noguchi and mollifying each other. Part of the problem came from their being prohibited from speaking Japanese in the laboratory. They claimed that although it was not permitted to speak anything but English at the institute, there was a good deal of German spoken in the dining rooms when the director was not around. But Noguchi was firm in enforcing the use of English in his laboratory.

One episode grew out of someone's report to Flexner that the young Japanese were not coming to the laboratory on time in the morning. Flexner spoke to Noguchi who told the students, "You must come in earlier." They took exception to this statement, but decided it was fruitless to argue in a language that was so difficult for them. Instead, they invited Noguchi to meet them for lunch on a Saturday at the Nippon Club where they could discuss the matter in Japanese. The issue focused on the word "must."

Noguchi explained that in using that word he was not "ordering them around" but that as guests of a foreign institute they were bound to conform to the local rules and customs. These Saturday lunches occurred from then on, helping the young Japanese to resolve their problems with Noguchi.

They found him to be just as hard on others as he was on the Japanese; they spoke of his screaming and cursing at his laboratory boy, Steve Molinscek, who took it in good humor. After Noguchi left the room, Steve would laugh and say that the professor had threatened to kill him again. Noguchi was equally arrogant in his attitude toward other scientists. So-and-so is third rate, he would say; someone else was second rate. Ohira got the impression that Noguchi regarded himself as the only first-rater. Yet Ohira mellowed in writing about Noguchi after his death, saying that he had been deeply impressed by the enormous prestige of the institute in New York and by Noguchi's having made it possible for Ohira to work there at a time when applicants were numbered in the hundreds. He remembered the day of his departure for Japan when Noguchi came running down the station platform, throwing magazines into his place on the train, and wringing his hand in farewell.[5]

Kudo was a little different, at least for a while. He was a zoologist who had already received his degree. He specialized in protozoa and was interested in diseases of silkworms and the plants on which the worms fed. Noguchi was frequently drawn to the study of parasitic protozoa. He had no experience with these organisms in the past and he knew that some workers considered that spirochetes were closer to protozoa than to bacteria.[6] As he moved from one thing to another in his search for something big, Noguchi attached himself for a while to Kudo's silkworm investigations.

The only mention that Noguchi ever made again of the Japanese academic community that is available to us came in April 1916, when he wrote Kobayashi of the students who came to work with him: "The world is a strange place—Japanese University Professors come to study under a person like me who has no connection with any group in Japan."[7] It is likely that he wrote a good deal more of his ideas on this subject to Chiwaki, but these details are lost.

Noguchi had started work by himself on Rocky Mountain spotted fever because of its similarity to *tsutsugamushi,* one of the endemic diseases of Japan that he had mentioned in his letters from Tokyo, but he accomplished little. Since the Japanese had moved into his field of spirochetes, Noguchi wanted to solve this other problem before they did, or somehow find a new trail to blaze.

In addition to the students and the papers that had begun to appear so suddenly from Japan, material containing spirochetes of Weil's disease arrived, addressed to Flexner but turned over to Noguchi to work with. After several attempts that failed because the spirochetes were dead on arrival, Flexner arranged for Dr. Miyajima, of the Kitasato Institute, to deliver a batch of inoculated guinea pigs to Dr. Victor Heiser in Tokyo. Heiser was making his first trip to New York in his new capacity as director for the Far East of the Rockefeller Foundation's International Health Board. It was from the guinea pigs brought to New York by Heiser that the strain was recovered and cultivated in Noguchi's laboratory.[8]

Heiser wrote amusingly of his problems with this project; not only did he have to contend with the objections of the ship's captain and the conductor on the train to the danger of having these infected animals on board, but he also had to make transfers of the organism to healthy guinea pigs he had brought along for the purpose. Two pigs were met at the station in New York by the institute's new automobile ambulance and were transported under full siren to the institute.[9] One of the guinea pigs was already dead, but after numerous transfers, the organisms were recovered from the other.[10]

It is not certain that Noguchi welcomed all this activity, particularly since it committed him to work with Weil's disease spirochetes at a time when he was looking in other directions. Evelyn Tilden says that Noguchi wanted to work with this organism, but at least in its early stages the work did not receive his primary attention. He had planned to spend a month at Woods Hole during the summer of 1916, and to take Kudo and his laboratory boy, Steve, along with him. Although he was never able to take a real vacation from his work, Noguchi loved Woods Hole, and working there took some of the pressure off him. He had read of *Cristospira* and *Saprospira*, spirochetes that are commensal with oysters and clams, and he was anxious to investigate them at Woods Hole. He also took along silkworms because he had heard that there were mulberry plants in the area.

Only two days before Noguchi was due to leave for Woods Hole, the first batch of guinea pigs arrived from Japan and he had to stop and make inoculations and prepare for work to be done with the inoculated animals in his absence, even though he was pessimistic about achieving success. (Nothing was recovered from this batch of guinea pigs.) About the same time a letter came from Flexner, who was in Seal Harbor, Maine, on his vacation, adding further to the anxiety of Noguchi's departure. There was a severe epidemic of poliomyelitis in the summer of 1916, and Flexner had written to urge Noguchi to involve himself in this problem in whatever way he might suggest.[11]

Noguchi had stayed clear of poliomyelitis ever since his first cultivation experiments in which he felt that he had been robbed of credit, and he had no intention of getting caught in this group frenzy now. He wrote Flexner a

four-page letter about the arrival of the guinea pigs and the arrangements
he had made for the work to go forward in his absence. When Noguchi
finally got around to mentioning poliomyelitis, he said he could not do
anything at the moment because of the mess in the laboratories from
building alterations that were in progress and he went off on a diatribe on
that subject: "It is indescribably dusty and even with the *closed trance and
door* the room is filled with dusts. No man can work with all these working
men hammering around. My room is the worst of all temporary quarters
furnished by Mr. James [the business manager of the Institute]—narrow
and near the parts extensively undergoing alteration. I am disgusted. I
have stood as long as I could. Until I am given a workable room I shall not
attempt to do anything. . . ."[12] Noguchi said there would still be cases of
poliomyelitis when he returned in September, and he would write from
Woods Hole what he wished to do. But he never did anything.

Many people had trouble with Mr. James, who left to join the army the
next year, and Noguchi was to have further problems. James fired
Noguchi's secretary while Noguchi was away at Woods Hole, and whereas
that led to a protest at the time, it led also to a positive development to
which we return later.

After an anxious settling-down period at Woods Hole, during which he
wrote Flexner endless details of the necessity for spending some $200 for
facilities there, Noguchi calmed down and enjoyed the time and work.
"Besides the laboratory work I have so much to do here. I brought my spotted fever slides (many thousands), unfinished Mss., some books on Trop.
Diseases, Protozoa, etc. Altogether it is very plesant to work here."[13] He
even decided that conditions would probably be all right in the
laboratories in New York by the time he returned. His little group identified a number of nonpathogenic spirochetes in the clams and oysters, but
never succeeded in culturing any of them. He was having such a good time,
finally, that he did not get around to work with the spotted fever slides or
the manuscripts.

Through the following fall and winter, Flexner seemed to be expanding
the influence of the institute in every possible direction. His enthusiasm
about the poliomyelitis work had been revived and he was pushing it hard;
in spite of Noguchi's having stayed clear of the actual work, he was drawn
in for public relations. Dr. John R. Paul, who devoted most of his professional life to work with poliomyelitis, was very critical of the institute's
work in that field. Of this particular period, Paul wrote:

> The author, then a second year medical student, can remember attend-
> ing this meeting during the Christmas vacation, 1916-17. Dr. Flexner
> was in the chair, and the several papers that were read must have been
> carefully selected to prevent the discussion from getting out of hand.
> Noguchi spoke of work going on at the Rockefeller Institute on globoid

bodies, and, as if to counter them, E. C. Rosenow of the Mayo Clinic described his findings on the bacterium which he thought was the responsible agent in poliomyelitis—a special form of streptococcus. Rosenow's presentation was received with scorn. . . .But I cannot recall any immediate skepticism being expressed about the globoid bodies, which were eventually to meet an even earlier oblivion than Rosenow's streptococcus. As far as the globoid bodies were concerned, it was apparent that Dr. Flexner, like Rosenow, was after something big. . .The manner in which Noguchi was introduced at the meeting invested him with an aura of mystery, as a mastermind who was about to clinch a great and useful discovery, or so it seemed to me.[14]

Beyond all this, World War I had been in progress in Europe for almost three years with the United States actively supporting the alliance against Germany. The feeling of war was clearly in the air and Flexner was just as anxious that the institute be seen to make an active, patriotic effort as he had been to join the war on poliomyelitis. Noguchi had recovered the Weil's disease organism from the Japanese material and Flexner arranged to have the European strain inoculated into rats and brought back from France by Carrel. Noguchi's work with Weil's disease was encouraged by Flexner because the disease was now spreading in the trenches in Europe. But it was only one of the many projects that were undertaken by the institute staff as a direct result of the war.

Carrel had been in France when the war started there. With the support of the Rockefeller Foundation, he began to study methods for the prevention of infection in wounds in a hospital at Compiègne. Carrel happened to be available to play nursemaid to the French rats because he was returning for consultation with the foundation and with the U.S. government. Flexner went with Carrel to Washington to seek American help for the French field hospitals, and as a result of their discussion, Flexner was encouraged to design a project for teaching and research for army physicians. His program was ready to receive army physicians for training shortly after the United States entry into the war. The enthusiasm was so great that the institute was designated a military hospital and laboratory, a flag was run up out in front, and all of the staff who were eligible were commissioned as officers in the U.S. Army. Dr. Welch became a colonel, Flexner started out as a major but by the end of the war he, too, was a colonel, but none of this was possible for Noguchi.[15] He did some teaching, but he worked mostly on the studies of Weil's disease—without even a lieutenant's uniform.

Noguchi established the European strain of Weil's disease from the rats delivered by Carrel, and he located a strain in rats in New York City. He now worked with some vigor on comparative studies of these two strains with the Japanese strain and demonstrated that all three were, indeed, different strains of a disease produced by the spirochete that had been identified in Japan, which had now been designated *Spirochaeta ictero-*

hemorrhagiae by Inada and Ido, the Japanese workers who had identified it.[16]

This was Noguchi's first real opportunity for a comprehensive study of a disease-producing organism that could be fully investigated with the techniques available at the time. Unlike syphilis, poliomyelitis, and rabies, a pure strain of the disease could be cultivated in the laboratory, and there was a susceptible animal, the guinea pig, that was cheap and relatively convenient for laboratory experiments. Although some of the spirochetes of Weil's disease were small enough to pass through coarse- and medium-grade filters, with proper staining they could be made visible under high-powered microscopes equipped for dark field illumination. Noguchi, without question, possessed unusual ability to use such techniques and he took advantage of this opportunity to enlarge his experience and gain an intimate knowledge of this particular organism, *S. icterohermorrhagiae*.

Noguchi was especially interested in techniques for identifying organisms by evidence derived from their immunologic reactions. When a susceptible human being or other animal is infected with a certain disease organism, antibodies are produced to attack that organism. After recovery from the illness, the individual retains these antibodies in the bloodstream and thereafter resists the disease. This is active immunity as opposed to passive immunity, which is temporary and is produced by injection into an individual of serum containing antibodies formed in another. If a passively immunized guinea pig resists infection when injected with a virulent organism, whereas a control nonimmunized guinea pig succumbs to the disease, this may be taken as evidence that the organism is identical with the one from which the passively immunized guinea pig had received its protection.

A second reaction, known as the Pfeiffer phenomenon, was also used by Noguchi in these comparative studies of the three strains of infectious jaundice. Living organisms injected into the peritoneal cavity of guinea pigs that have been rendered immune undergo certain changes that can be observed when the peritoneal exudate is withdrawn and examined at various intervals. The organisms gradually become nonmotile, swollen, coarsely granular, and indistinct in outline, and finally disappear altogether; this lysis of the organisms is known as the Pfeiffer phenomenon. The Weil's disease organism could be identified by this method.

These two methods of identification became an important part of Noguchi's technical armamentarium. In fact, he was teased about his affection for guinea pigs since he always had so many on hand in the laboratory and later took so many with him when he traveled on scientific expeditions.

Beyond these immunologic studies, Noguchi particularly wanted to understand the structure of these organisms and to learn how they moved.

S. icterohemorrhagiae is a very tiny spirochete, and this fact seemed to challenge him. "Its elementary structure is that of a closely wound cylindrical thread. . . .The number of coils is greater in a given length than that of any spirochete heretofore known. . . . The winding is rarely seen distinctly, although it can be brought out well by a carefully stained preparation . . .or under darkfield illumination." He described the organism not only in detail but, contrasting its characteristics with other known spirochetes, he believed that differences justified the creation of a new genus; "on account of its fine and minute windings, the name *Leptospira* is suggested."[17] This new genus was readily accepted by other bacteriologists and is still recognized today. It now includes several other microorganisms that have been classified as *Leptospira*.

The first paper in which Noguchi reported his studies of Weil's disease was published in 1917, but publication of the comparative studies of the three strains was delayed until 1918 because his work was interrupted by a prolonged illness. He thought at first that he had contracted Weil's disease in the laboratory. While Noguchi was drawing some highly infectious material into a pipette, Peyton Rous stepped into the laboratory behind him, startling him into drawing some of the fluid into his mouth. Noguchi quickly rinsed his mouth with alcohol, and for a while thereafter he would ask his technician when he came into the laboratory in the morning, "Do I look yellow—more so than usual, that is?"[18] But one day Noguchi did not come to the laboratory at all. He had been taken to Mount Sinai Hospital, under the care of his trusted friend and physician, Dr. Libman, who diagnosed Noguchi's condition as typhoid fever rather than Weil's disease.

His wife and his Japanese friends reported that he had been eating large quantities of raw oysters on the theory that he might thus restore his old feeling of well-being. The Japanese students added that Noguchi would surely have fallen ill of something else had it not been for the oysters; he had been in such poor spirits for so long that they considered him to be in a highly vulnerable state.[19] And in addition, it was his forty-second year (by the Japanese way of counting age), a year of danger for Japanese men, *yakudoshi.* He spoke of it that way himself, in jest, after he had recovered.[20]

For the first several weeks of his illness Noguchi was critically ill; he had a severe case of typhoid fever from which he developed complications—a perforation of one of the visceral organs. Since Noguchi was known to have an enlarged heart, Dr. Libman was afraid to risk surgery, which is usually required to repair a perforation, so Noguchi suffered for weeks with high fever from the resulting peritonitis as well as the typhoid. He had been taken to the hospital on May 24, 1917, and on June 22 he was able, for the first time, to write to Flexner in a very shaky hand that he thought he would survive.[21]

Weak as he was, Noguchi wrote a second letter to Flexner on the same day, saying: "My technician Steven Molinscek has been more than efficient in his work ever since my illness. He kept every culture intact and going and kept the Rocky Mountain Spotted Fever viruses straight. When the tall boy was discharged he was increased by $10, making his wages $65 (or $70). You indicated to me that another $10 will be raised from July. This additional note is to remind you of this case as if he be left out of promotion I must feel almost criminal. He had a wife and baby."[22] He was fond of Steve; and Steve had shown his confidence in Noguchi's recovery by bringing culture tubes into the hospital to show them to Noguchi ever since he had been well enough to open his eyes.

Following his discharge on June 22, Noguchi suffered a relapse and was readmitted to the hospital. After his second discharge from Mount Sinai on August 1, he went to a resort hotel in the Catskills with his wife for recuperation. He arrived there about August 7 and seemed to be gaining in strength rapidly when he suffered a second relapse. He was brought back and admitted to the institute hospital on September 14, remaining there until October 27. But Noguchi's troubles were not yet over. He became ill again in December and was readmitted to Mount Sinai with a diagnosis of subacute appendicitis, and this time he underwent surgery. Mount Sinai records are no longer available, but physicians of today suggest that, since Dr. Libman found that Noguchi's blood count was normal and he had no fever, his problem may have been adhesions that had formed in the healing of the perforation. In any event, Noguchi was finally discharged from Mount Sinai on January 4, 1918, and went on to recovery and a return to his work.[23]

Between the time of Noguchi's last admission for typhoid and his admission for surgery, his wife was rushed to the hospital and operated on for acute appendicitis, so the Noguchi's medical expenses, including the stay of almost six weeks in the Catskills, had mounted up and again Noguchi appealed to Hoshi. The amount of money Hoshi sent was sufficient not only to allow Noguchi to pay the bills, but also to buy a used car and a lot at Shandaken, the village where he had stayed in the mountains. Noguchi had fallen in love with the area and had found a piece of land he wanted, land with a trout stream running through it. People said that it reminded him of the Aizu area of Japan, and perhaps the whole experience was something of a reversion to childhood things. The illness had forced Noguchi, for the first time in his life, to give up work and allow himself to be taken care of. His painter friend Hori had given him paints and brushes when he first went to Shandaken and he started teaching himself to paint, a hobby that he cultivated for the rest of his life. He also read books written for Japanese children, legends of the heroic characters of Japanese history.[24] He seems never to have read these stories in his own childhood, such stories as those of the wandering samurai Musashi Miyamoto, which most Japanese children know well. Noguchi was delighted to learn of these heroes for the first time, and seemed to find some of the childhood he had missed.

As soon as he was back at work, Noguchi took an active interest in the design and construction of a summer cottage on his land. In February, in the same letter in which he wrote of his illness to Kobayashi and his mother, Noguchi described the plans for the cottage.[25] But he was no more than out of the woods from his illness when a different kind of tragedy befell him.

On March 18, 1918, his laboratory assistant, Steve, fell ill, and died of Rocky Mountain Spotted Fever a week later. Steve was only twenty-three years old. At his death, he left a wife and a baby, and, perhaps worst of all for Noguchi, Steve knew that he had scratched himself with a needle he was using for injecting animals but did not report the accident. Noguchi felt that if it had been known sooner, Steve could have been given protective serum that would either have prevented the full development of the disease or modified its severity to the extent that he could have recovered. Why Steve failed to report the incident is unknown.

Noguchi reported his own investigation of the matter in the form of a handwritten letter to Flexner. He gave the results of a study of Steve's blood, of the clinical progress of the illness, and of injections of the blood into animals, Noguchi concluded:

> Now it appears that the patient had an accident in his performance of inoculation by sticking the charged needle into some part of his hand or arm. Since my last illness he was put in charge of carrying the virus in guinea pigs by periodic passages and was doing so independently during the last nine months until the accident took place presumably in early part of March this year. The work is now being taken into my own hands again and will remain so until the cessation of the investigation.
> I am sorry to report this incidence to you and shall accept any responsibility for what may have been my share of fault in the case.[26]

That was the end of it, but Noguchi's solution was typical of him. He would now take the matter back into his own hands, assuring safety to others by taking all the risks himself, perhaps even feeling that these risks were either nonexistent or greatly minimized in his hands. As with his relationship to the Japanese, Noguchi could not stand having trained people around him. He had to take completely untrained workers and make technicians out of them in his own way. He must have been desperately afraid of his own lack of knowledge showing up. Sincerely saddened though he was by Steve's death, Noguchi still could not admit that he might decrease the risks by using trained personnel.

It was this attitude that led, in part, to his making a technician out of the secretary who was assigned to him after James fired the previous girl. Evelyn Tilden had graduated in 1913 from Pembroke, the Women's College of Brown University, where she had majored in English literature and foreign languages (German and French). Her father, who had been a minister, died shortly after her graduation and she became the sole support of her mother. She went to work as a librarian, but to increase her earnings she changed to work as a hospital librarian. On the strength of her language training and small experience in a hospital, she was employed by the Rockefeller Institute after an interview with Miss Von der Osten, Dr. Flexner's secretary. For some time, however, Miss Tilden was ignored by Flexner.

Before Miss Tilden's arrival, Flexner had edited all of Noguchi's writing. When Noguchi learned of Miss Tilden's background, he asked her to edit the English in a current paper of his, and when this paper reached Flexner's desk he recognized at once that a new influence had been brought to bear on Noguchi. Flexner sent for her, expressed his delight, and encouraged her to continue. By the following spring (1917), Miss Tilden was already working as a technician.[27]

She was a large woman, five feet nine inches. She loved vigorous physical activity, and particularly hiking and mountain climbing. She had grown up in a small town in Maine where once, at the age of seventeen, she had thought of marriage, but her family had sent her off to boarding school to end that romantic episode. She says she got over the idea of marrying and was never tempted again. From the beginning of their association, Noguchi found something in Miss Tilden that he needed, and her devotion to him was complete.

He was soon encouraging her to take formal training in bacteriology, and he started her out on practical work almost immediately. "The cultivation of the mouth spirochetes was my first problem, when I became Noguchi's assistant, and my introduction to general bacteriology. One would hardly select this for a first exercise in bacteriology for a class, but Dr. Noguchi's method of teaching by beginning at any given point and working backwards or forwards seemed to work quite as well as any other."[28] This was a repetition of work in pure culture of a mouth spirochete, a new species, which Noguchi had published in 1912. He appears to have used this as a means for teaching his technicians his cultivation methods, as he had Ohira and others working on similar problems.

Miss Tilden related this to work that Noguchi was doing at the time on the organisms causing the relapsing fevers, one of which he had cultivated in 1913. (It was the one that had been described by Obermeier in 1873 and was the organism Noguchi had seen under Watanabe's microscope in Wakamatsu). Noguchi and Ohira published a paper in 1917 on the cultivation methods that Noguchi had developed in working with the syphilis spirochete.[29]

Miss Tilden studied by herself and finally enrolled in courses at Columbia where she earned her master's degree in 1926. She spent two summers working at Woods Hole taking advanced courses for which she was most inadequately prepared, but she managed to fill in background as she went along. An essential part of her relationship with Noguchi came from her need to learn and her ability to undertake formal course work for the education he needed but which his pride made impossible. Noguchi could have done these things in Philadelphia, but in those days his lack of money and his language problem had prevented him from doing so. Miss Tilden taught him as she learned, and the fact that he could accept it from her was full of meaning for them both.

She continued working toward a Ph.D., which she says both Noguchi and Flexner encouraged her to do. At the time of Noguchi's death, she had not finished her graduate work but Dr. Flexner kept her on at the institute to finish Noguchi's work, and with Flexner's encouragement she managed to complete the work for her doctor's degree as well. During this time Eckstein was working on his biography of Noguchi. Flexner not only refused to see Eckstein himself, but he forbade Miss Tilden to do so. But when Eckstein called her at home and told her that he was going to do the biography with or without her help, she thought that she would rather have some control over the biography than give Eckstein free rein by refusing her help. "Sometimes Dr. Flexner had to be overruled." She said that Eckstein was very nice to her and that he kept his promise to keep her name out of the biography and to omit material that she considered too personal and "nobody's business." Although Flexner was critical of the book, Miss Tilden noted that he gave copies to several people as gifts.

Miss Tilden had worked with Noguchi for twelve years at the time of his death, and worked on his projects for three years thereafter. She wanted to write about him but was never able to do so. That Noguchi determined the course of her life goes without saying. She became a full professor at Northwestern University School of Medicine, and regrets that neither Noguchi nor Flexner could know this as they had both given her so much help. What she meant to Noguchi was less obvious; he liked big women and got some assurance from their strength. Miss Tilden tells of his measuring her height and his own one day and then saying, "Well, what's seven inches!" She claims that Noguchi was much more sensitive about his height than he was about his hand, an observation no one else has made.[30]

Although his summer cottage at Shandaken meant so much to Noguchi, and Miss Tilden was very fond of the mountain environment, she never visited Shandaken. This would certainly be the result of a desire, perhaps on everyone's part, to keep her and Mrs. Noguchi apart. Miss Tilden claims they never met.

By the time Noguchi returned to work after his illness, the institute was in full swing with its participation in the war work, and although he did some teaching, he did it as a civilian, which was not at all to his liking. Again he was isolated, and again he was irritable. Baron Matsumoto's son was working with him on a rather dull problem, and Dr. Kanematsu Sugiura feels that Noguchi had a special obligation to help this young man since he had been helped by the young man's father in his own early days in Tokyo. Yet, after the son had worked for a year and gotten nowhere with his problem, he spoke to Noguchi about it and Noguchi said, "OK, then, you might as well get out." Young Matsumoto was horrified by such treatment, and told his tale to Sugiura, who got in touch with Dr. Kolmer at the University of Pennsylvania. Kolmer took Matsumoto on. They did an acceptable research project, and Matsumoto returned to Japan and got his degree from a Japanese university.[31] Miss Tilden says this was just the procedure that Noguchi found intolerable. The old Japanese idea that only Japanese university degrees were acceptable, even though students had to come to the West to do their research, infuriated him and he completely refused to cooperate.

But he was saved from further irritation and floundering by an appointment by the Rockefeller Foundation of a commission to study yellow fever in Ecuador. Noguchi was assigned to that commission as a bacteriologist, and he wrote to Kobayashi from shipboard on his way to Panama: "I was requested by the American Army and the International Health Board to go to Ecuador in South America to work on yellow fever after I recovered from my illness. . . . Everyone feels it is dangerous to visit the tropical areas but I think it is not that bad. . . ."[32]

16

Yellow Fever in Guayaquil

As for myself the whole affair was and still is a model of harmony.
— Noguchi to Flexner, September 18, 1918[1]

YELLOW FEVER PRESENTED JUST THE KIND OF INTERNATIONAL PUBLIC
health problem that the Rockefeller Foundation was most eager to attack.
The disease had a spectacular history. For Noguchi, the opportunity to
join an all-out effort to eradicate the disease was a lifeline on which to pull
himself out of his prolonged depression; it was the new goal he had been
seeking.

From obscure origins in Africa, yellow fever had been carried to tropical
and subtropical port cities of North and South America in the seventeenth
century. There it had grown into an endemic urban disease, bursting
regularly into epidemics that ravaged cities as far north as New York and
Boston. Philadelphia, Charleston, New Orleans, Galveston, and Havana
had all learned to dread the epidemics. The names of famous men
associated with past efforts to halt the disease or to discover its cause were
enough in themselves to attract such an ambitious nature as Noguchi's.

Following the Spanish-American War, when as many soldiers died of
yellow fever as in battle, scientific investigation yielded the first clues.
Together with Cuban physicians, Major Walter Reed and his co-workers in
the U.S. Army demonstrated that the disease was carried by a mosquito.
Since they could not find a susceptible animal for experimental work, the
drama of their demonstration was heightened by their use of human
volunteers. Following this demonstration, Major William Gorgas, the chief
sanitary officer in Havana, directed a mosquito eradication program in

Havana that showed that the disease could be brought under control by this public health measure. Gorgas repeated the mosquito eradication program in Panama, enabling the Army Corps of Engineers to construct the Panama Canal after previous efforts by the French had been defeated by yellow fever and malaria. Once the canal was open to shipping in 1912, the potential for the expansion of commerce and industry stimulated an interest in improving public health and safety in Central and South American cities that were still frequently subject to quarantine. The newly created Rockefeller Foundation selected the elimination of yellow fever as its first major endeavor in the interests of the betterment of mankind.

In 1916, General Gorgas, who had become U.S. Surgeon General, served as chairman of a commission that spent six months surveying yellow fever in South America for the foundation. The commission identified Guayaquil, Ecuador, as an endemic center of yellow fever, and recommended that a mosquito eradication program be initiated in that area. World War I intervened, but when Gorgas was freed of his army duties in 1918, he took charge of the Rockefeller Foundation yellow fever program. He started with the appointment of a special commission to visit Guayaquil, nominally to gather more up-to-date information about the problems there, but also to learn whether a mosquito eradication team from the United States would be welcome in Ecuador. To many people in that country, Yankees were just an aggressive young version of the Spanish imperialists whom they had thrown out almost a hundred years before. They regarded the designation of Guayaquil as a primary endemic center for tropical diseases as a gratuitous insult. The frequent quarantines of the city had already created hardships, and its citizens were neither trusting nor enthusiastic.

Gorgas chose his small commission to Guayaquil with care: as chairman, Arthur Kendall, pathologist, dean of Northwestern University School of Medicine, a man of tact and enthusiasm who had worked with Gorgas in Panama; as clinician, Charles Elliott, and as chemist, Herman Redenbaugh, both also of Northwestern; as yellow fever specialist, Mario Lebredo of Havana; and as bacteriologist, Noguchi, whose recent work with the organism of Weil's disease commended him to Gorgas.[2]

The original Cuban studies had shown that yellow fever was carried by the mosquito *Aedes aegypti*, and that the organism passed through all known filters and was thus, by the definition of the times, a virus. They had further demonstrated that the causative agent was found in the bloodstream only for three or four days after the onset of the illness. The organism had never been identified, but Gorgas was among those who looked with favor on the prediction that one day this organism would be found to be a spirochete. Flexner, an active member of the Board of Directors of the foundation throughout the time of planning the yellow fever program, may have held a similar view. During this time Flexner had been pushing forward the investigations of Weil's disease: he had brought the papers reporting the discovery of the spirochete of that disease from Japan and published them in the journal of which he was editor; he had arranged for the transportation of the organisms from Japan in the guinea pigs attended by Dr. Heiser, and from France in the rats similarly attended by the illustrious Alexis Carrel.

Although the objective of the Guayaquil commission was to set the scene for a mosquito eradication team to follow, a hunch was nourished, in hope and ignorance, that yellow fever and Weil's disease might be closely related and that a bacteriologist who was familiar with spirochetes might be able to learn the nature of their relationship. The Japanese workers who had found the Weil's disease organism had only textbook descriptions of yellow fever. Drawing on these descriptions, they noted that the only difference between the two that was apparent to them was the transmission of yellow fever by mosquitoes, there being no evidence that Weil's disease was ever transmitted in this way.[3] On the American continents there had been almost no clinical experience with Weil's disease, and there was a similar lack of understanding of the distinguishing features of the two diseases. Noguchi had no clinical experience with either disease; he was chosen for the commission on the basis of his experience with spirochetes in the laboratory.

When Noguchi arrived in Guayaquil on July 15, 1918, he felt that he had a specific assignment, detached from whatever the other members of the commission might be doing. He had traveled alone from New York a week after the others, having been delayed by visa problems related to his Japanese citizenship.[4] He had to change ships in Panama, and when he discovered that his seventeen crates of equipment and supplies were not on board the ship bound for Guayaquil, he created enough of a disturbance that a special train was sent from Colón to deliver the crates to the ship at Balboa, "just in sheer time."[5] Even at that, two crates were missing, adding to the flurry of his arrival at Guayaquil. Kendall wrote: "Noguchi arrived Monday and started at once in the laboratory, buying material to enable him to do so. His enthusiasm is tremendous and so far I have been utterly unable to keep up with his wants. . . .Noguchi orders ad lib and I get all sorts of statements, verbal and otherwise, which keeps me in turmoil and my financial status in chaos."[6]

Noguchi was delighted with his reception, with the fine living quarters assigned to him in a hotel "which is spacious, screened and up-to-date in every respect," and with the cooperation he received in getting his work started. He wrote to Flexner: "Dr. Kendall is making a general survey of the sanitary situation with the officials of the city. Apparently he had been appointed President of the Commission by the Rockefeller Foundation, although I have never been told of this fact by Dr. Rose* or anyone else."[7] If Noguchi had been informed of the broad objectives of the commission, he never let the knowledge distract him from his own pursuits.

Cases diagnosed as yellow fever were plentiful, and the day following his arrival Noguchi transferred blood from patients to the guinea pigs that he had brought with him. On August 17, barely a month after his arrival, Noguchi wrote Flexner of the well-ordered and rational program he had followed, and reported that he had already found a spirochete. He had brought Weil's disease immune serum with him for testing purposes, and told Flexner that "the immune serums (three different samples) prepared by me at the Rockefeller Institute also protect the guinea pigs against the strain obtained here." Noguchi said that he had produced a disease in guinea pigs with his Guayaquil spirochete that he said was a *Leptospira*, the genus he had created for the spirochetes of Weil's disease, but he maintained that it had come from cases of yellow fever and that the disease it had produced in the guinea pigs showed all the symptoms of yellow fever in humans.[8]

It is frequently mentioned that Lebredo, the Cuban yellow fever specialist of the commission, talked with Noguchi about his work as it progressed in Guayaquil and that he warned Noguchi against too easy acceptance of the diagnoses of the cases presented as yellow fever.[9] Lebredo was worried because many of the patients he saw in the hospital had two or more of the several tropical diseases, and also because of Noguchi's belief that he had produced yellow fever in animals.[10] The use of human volunteers in the demonstrations in Cuba had made all physicians from that country especially aware of the failure in the past to transmit the disease to any animal. It is not clear just when, or how explicitly, Lebredo expressed his concerns to Noguchi. But there is no indication that Noguchi gave much thought to the possibility that there could be two separate diseases that were, perhaps, not being correctly differentiated in diagnosis and further, that both these diseases might coexist in the same patient. On the contrary, Noguchi wrote at this time as though he believed that yellow fever was the local manifestation of Weil's disease: "One feature which seems to differentiate the Guayaquil strain from those of the Japanese, American and European is that the organism disappears from the blood and tissues before the death of the guinea pigs. . . .The Guayaquil strain. . .also tends to produce more pronounced icterus and liver changes with less hemorrhagic lesions."[11] He started some experiments in which he tried to transmit the disease by means of mosquito carriers even as he noted that a large percentage of the poorer classes of people wore no

---

* Wicklife Rose was director of the International Health Commission, Rockefeller Foundation, 1913-16, and director of its successor, International Health Board, 1916-23.

shoes, which would have exposed them to the same means of infection as had been established for Weil's disease in Japan and Europe.

He commented briefly on the work of other members of the commission and said: "Their work is almost finished and [they] will be ready to return early in September. . . .I hope to finish most of my work by the time when they start here, but no one can tell what may come up yet to compel my stay a little longer. I am wondering what will be the attitude of Dr. Rose in regard to my further stay here in case of advisability."[12]

Dr. Lebredo's report of his work in Ecuador, written in Spanish and condensed for the final joint report in collaboration with Dr. Kendall, strongly emphasized the complicated epidemiologic situation in Guayaquil. Lebredo showed particular concern over the impossibility of accurate diagnosis, and of the need for early recognition of yellow fever and bubonic plague. He said nothing of Weil's disease, but he emphasized the inadequacy of facilities for isolating cases of infectious disease in the general hospital. He said that almost all of the patients who were admitted to the yellow fever hospital were critically ill, since diagnoses were not made before symptoms were well advanced.[13]

It seems that Lebredo's warning may well have been perceived in the terms that had been so insulting and caused such hard feelings among the Ecuadorians in the past. Perhaps there was some rivalry in the air—Dr. Guiteras of Cuba had been a member of the commission that had given Guayaquil the reputation of the filthiest city of the Western hemisphere. Both Kendall and Noguchi wrote with pleasure and surprise of their first impressions of Guayaquil, and there is even one story of a government official appearing in Kendall's hotel room to prostrate himself with thanks for Kendall's kind words about the city shortly after his arrival.[14]

The local physicians were just as sure that they could recognize yellow fever when they saw it as Lebredo was concerned about the public health and hospital conditions as he saw them. Noguchi was not the only member of the commission to ignore Lebredo's warnings and to accept what the Guayaquil physicians believed about yellow fever cases. In his final report, Elliott, the clinician, stated:

While the mosquito. . .has been considered the chief, if not the only, means of transmission of yellow fever, in view of the similarity that yellow fever appears to bear to infectious jaundice, other means of transmission, such as the transdermic route as demonstrated in the latter disease, seem possible although unusual. The fact that many of these patients went barefooted habitually and were admitted with many infected sores on the exposed parts of the body would lend weight to this possibility. . . .Yellow fever as seen in Guayaquil does not appear to differ in any essential feature from infectious jaundice as reported from Japan and Flanders, such differences as there are appear to be differences in degree rather than in kind, a degree which does not appear to be greater than that found in the clinical descriptions of yellow fever from widely separated areas.[15]

Noguchi's plans for further extension of his work at this time hint that he was already thinking in terms of solving the whole question of the etiology of yellow fever in these few months. When he published reports of the work he had done during this period in Guayaquil, Noguchi made numerous references to the pressure of time, indicating that he would have assembled more data if time had permitted. Why did he feel that he had to cover so much ground in such a short time? Realistically, the only explanation could have been that the planned mosquito eradication program would wipe out the disease. For one whose interest lay in finding the causative agent, whether for its scientific value or for his own ambition, the elimination of the disease would certainly be a complication. Although most workers would assume that the disease would undoubtedly appear in some other place, Noguchi could not wait for such unpredictable opportunities. He was always so excited by even the thought of getting something he wanted that his greed became boundless and, inevitably, his professional detachment disappeared.

With all his work, he was having a wonderful time. The Ecuadorians were generous and friendly and Noguchi found their ebullient admiration for his work warming and seductive. In response to their respectful interest as well as his own dreams of glory, he was expanding the scope of his work further and further. By the time of his next letter, on September 2, Noguchi thought that some of his mosquito transmission might be successful.[16] He had obtained a pure culture of one strain and was proceeding to prepare immune serum. He had definitely decided to remain a little longer, Kendall had approved the decision and left Noguchi with the necessary funds as the rest of the commission departed on September 3.

In spite of the problems that had been anticipated, Kendall had been reasonably successful in negotiating an agreement for joint responsibility between the Rockefeller Foundation and the government of Ecuador for the mosquito eradication program.[17] At the last moment Kendall was thwarted in getting a signed agreement by the ill-timed announcement of new restrictions on the amount of cacao the United States would allow to be imported; these restrictions were offensive and the president of Ecuador had to protest. For this reason, he could not sign the agreement, but he gave the plan his verbal support and promised the required cooperation. All this took a great expenditure of time and energy on Kendall's part; he had already written to Rose at the foundation that he had been so fully occupied that he was unaware of the details of the work of other members of the commission.[18]

But Noguchi was just as unaware of Kendall's problems. Noguchi wrote of having shown his work freely to all who were interested, and he complained that Kendall was the only one who was unfamiliar with his work.[19] After his return to Chicago, but while Noguchi was still in Guayaquil, Kendall notified Rose that a sanitary survey of the Republic of Ecuador, by himself and Lebredo, had already been mailed and he listed the other reports that were to follow: a clinical study by Elliott with a supplementary clinical report by Lebredo were among the items listed.[20] When the actual reports were submitted, however, the list was changed. A report on medical education in Ecuador had been added, but Lebredo's supplementary clinical report was not included.[21] Again it would seem that the need to protect the feelings of the sensitive citizens of Guayaquil was a consideration.

Kendall's decision to add a report on medical education in Ecuador reflected observations that led him to the opinion that

> great possibilities may flow from the visit of the Yellow Fever Commission to Ecuador. The Ecuadorians feel keenly the bondage under which they have labored through cut-throat concessions. . . .They are also keenly alive to the impression somewhat generally held by the Anglo-Saxon race that the Latin Americans are untrustworthy, of doubtful antecedents, and of swarthy color.

Kendall believed that educational and humanitarian expeditions held enormous potential for overcoming the previous blunders of foreigners in Ecuador, and he expressed glowing respect for the people whom he had met in that country.[22]

With Kendall so enraptured by the persuasive charm of the Latin Americans, how much resistance could be expected from Noguchi who could so easily identify with their anti-American feeling? Left on his own in Ecuador, and basking in his work, Noguchi ventured still further in his efforts to make a great discovery for himself and for these people he was finding so responsive. On receipt of Noguchi's optimistic letter of August 17, Flexner had cabled him not to make any public announcement of his findings until after his return to New York.[23] Noguchi answered this with a long report of the cordiality of the local people:

> They were intensely interested in my work and spared no effort to promote it. They were generous and openhearted and this was reciprocated by me in allowing them to follow the evolution of work step by step. By this way I was fortunate to have their confidence and enthusiasm. . . .I spared no effort to explain my findings when Dr. Pareja or any other wanted to know. There was no secret in the whole work and I of course asked them to keep things from the public and they did. The governing class of Ecuador is relatives of each other and the findings in the laboratory or hospital were an open secret — that is — everyone of any importance was gradually informed of the advents of the investigations. No public announcement has been made, but they know just the same.[24]

This letter was written from Quito where Noguchi met the president and members of the Ministry and where he had gone to inoculate some seven hundred nonimmune soldiers coming from the highlands to Guayaquil for the national Independence Day celebrations. He said that the government officials understood that this was only a preliminary experiment, claiming that he had still made no public announcement.

Noguchi's reception by the officials in Quito was only the beginning of a month of banquets, awards, and citations, which, in their very special meaning for him, are best described in a letter Noguchi wrote to Madsen the day before his departure for the United States. Before writing of the celebrations he told of his illness in New York—"All hopes for recovery were at one time abandoned"—and of his whole experience in Ecuador. Noguchi said that he had transmitted yellow fever to animals, had found and cultivated the virus, studied immunity phenomena, and performed some prophylactic vaccinations:

> The organism was found in the blood and tissue of yellow fever patients and in the experimental animals. It is visible under the darkfield microscope and is *filterable*. Its form is a leptospira, similar to the organism of Inada and Ido who found it in Weil's disease. But immunity reactions differentiate the two. The pathogenic properties of these two organisms are also different.[25]

It is likely that Noguchi had known, before he left New York for Ecuador, that on February 20, 1918, a paper had been received for publication by the *Journal of Experimental Medicine* from Japanese workers at Kyūshū—Ido, Ito, and Wani. These men had discovered a spirochete that was very similar to the spirochete of Weil's disease and was the cause of seven-day fever (*nanukayami*), a disease that had been recognized only in Japan at that time. The disease was similar to the milder forms of Weil's disease, but the Japanese workers had found that the organism could be distinguished serologically from the spirochete of Weil's disease. They gave it the name *Spirochaeta hebdomadis*, a designation that corresponded to the Japanese designation *Spirochaeta nanukayami*, which they had used for the original presentation in Japanese journals.[26] It is not possible to know, of course, how much his knowledge of this discovery affected Noguchi's interpretation of his findings in Guayaquil. But within a little more than two months after his letter of August 17 to Flexner, Noguchi had changed his views; his new interpretation of his findings, as he wrote it to Madsen, corresponded exactly to the observations of the Japanese workers of the organism of seven-day fever. Noguchi appears to have ignored the facts that the Japanese workers could be sure of the diagnoses of the diseases with which they were working and that their findings in their serologic studies were consistent, whereas his findings were more variable than otherwise.

In this letter to Madsen Noguchi also noted that "Leptospira is a new

Hideyo Noguchi, officer in the Army of Ecuador, 1919. Hideyo Noguchi Memorial Association

genus created by me recently." He now believed that yellow fever was not the same as Weil's disease. The organisms of the two diseases were similar; they were both *Leptospira,* but they were different, and he could demonstrate the differentiating features. Having thus stated the arguments through which he had convinced himself that he had discovered the yellow fever organism, Noguchi boasted a little to Madsen of what he could not have confided to anyone else:

> My work has been closely followed by the medical men here and they are so well satisfied with the findings that they all consider that the cause has been found. . . .As you may see and amuse yourself your first foreign pupil has been made "cirujano-major and Coronel" of the Ecuadorian Army, with a present of sword! Now when I go back I may teach these U.S. Medical Officers in my own uniform, because in the U.S. no foreigner is entitled to an Army rank and I was teaching these men as a private person! At the same time I was given the freedom of the City of Guayaquil and of Ecuador—citizenship passed by the Congress here.[27]

The awards and banquets and honors were even more elaborate than his letter indicates. The Public Health Service announced that a bronze tablet would be cast and hung in the hospital where he had worked in Guayaquil, inscribed as follows: "In this laboratory of the Public Health Service the eminent Japanese bacteriologist, Hideyo Noguchi, Member of the Rockefeller Institute, discovered the germ of yellow fever on the 24'th of July, 1918."[28] Phrases such as "cradle of a discovery so important both for humanity and for science" abounded, and before Noguchi wrote his letter to Madsen, at least ten organizations passed lavishly worded resolutions. Copies of all the awards and citations were sent to the Rockefeller Institute and to the government of Japan.

In this atmosphere, just too seductive for him to resist, Noguchi had committed himself to the interpretation of his findings which he communicated to Madsen: that there are two diseases, produced by two similar but not identical *Leptospira,* which can be differentiated from each other by their pathogenic properties* and by their immunity reactions. There is no record of the evidence on which he based these conclusions, but whatever it was, even Flexner found it inadequate when Noguchi presented it back in New York.

In his written report to the foundation after his return, Noguchi stated that he had found an organism but that further immunologic work would be reported later.[29] But the whole burden of Noguchi's claims that his new

---

Pathogenicity is a term that has often been used incorrectly in place of the term virulence. Pathogenicity denotes a general capability to produce disease and may refer to agents other than microorganisms. A specific agent is either pathogenic or it is not. Virulence is the property only of a microorganism and introduces the concept of degree. It is a characteristic of a particular strain, and is the measure of the severity of the disease produced. In this instance, as in many others, Noguchi used the term *pathogenicity* incorrectly.

Leptospira was the cause of yellow fever fell on the differentiation of this organism from the *Leptospira* of Weil's disease by means of immunologic reactions of the two. Apparently ignoring his own observation of Weil's disease in wild rodents in Ecuador (which he published some time later[30]), ignoring the warnings of Lebredo against diagnostic confusion, ignoring his own previous finding that the Weil's disease immune serum he had taken to Ecuador from his laboratory in New York gave "protection against the strain here," and ignoring the irregularity with which he was able to achieve his results, Noguchi created a rationalization that met his own needs, that catered to the pride of his new friends, but met no scientific criteria.

True to his old style, Noguchi had touched on numerous aspects of a very complicated problem, collecting only a small body of data on each aspect, and having been successful in communicating the disease to animals from only six cases of a total of twenty-seven supposed cases of yellow fever with which he had worked.[31] Yet his whole reputation had been built on this style, and he had never been called to account for its deficiencies. In truth, Noguchi had accomplished much and his power to persuade was seemingly irresistible. Rabies had been forgotten, but it was still considered that he had been successful in the cultivation of the syphilis spirochete. In spite of his having found syphilis spirochetes in only a few of the paresis specimens he had examined, it turned out that he was correct. His poliomyelitis cultures did transmit the disease to susceptible animals even though no one knew enough about viruses to combat Flexner's globoid body interpretation effectively. His work with Weil's disease had been consistent and productive, as had the cultivation work with many other spirochetes.

As the results of each step of his work caught him in an upward spiral to even more dramatic steps rather than directing him to cautious verification of his findings and careful consideration of their meaning, Noguchi's judgment became more and more subject to his heady dreams and finally vanished. The enthusiasm of his reception in South America further obscured his caution. He was well acquainted with such feelings of denigration as the South Americans had experienced. Perhaps he had hinted that he understood, and it was the hinting that led to their orgy of awards. They somehow came to realize how acceptable these expressions would be to him, in contrast to some of the Anglo-Saxons who were less vulnerable to such flattery. At the time of his letter to Madsen, only Noguchi knew of the fateful decision he had made, oblivious to its destructive potential even for himself. He was totally immersed in his immediate sense of conquest and the adulation of "the people of importance" in Ecuador. Noguchi sailed on October 27 in a happy glow, and by the time he arrived in New York on November 24, he had written no fewer than nine papers on the subject of his four months' work in Ecuador.[32]

No one was able to state, with the assurance of those who read his papers today in the light of the experience of the intervening years, that Noguchi's descriptions of the cases he saw did not correspond to a true picture of yellow fever.[33] Lebredo had said so, but this implied a criticism of the clinicians of Guayaquil. And since he also spoke with unpopular frankness about the sanitary conditions of that city, his observations were not welcome. The satisfaction that Noguchi was able to give, through experiments performed during the first few months after his return, on the seemingly tiny matter of the immunologic differences between the so-called *Leptospira,* led to the announcement of *Leptospira icteroides* (as he later called it):[34] the organism he had created for reasons that allegedly are never present in scientific experiments; the nonexistent organism that has since been categorized as the red herring drawn across the whole field of yellow fever research for ten years.[35]

17

Mexico, Peru, and the Fatal Commitment

FLEXNER WAS IN EUROPE WHEN NOGUCHI RETURNED TO NEW YORK ON November 24, but a special group that included General Gorgas, Mr. Rose, Dr. Welch, Dr. Theobald Smith, Dr. T. Mitchell Prudden,* Dr. Heiser, and Dr. Rufus Cole assembled to view the results of his yellow fever studies in Ecuador. The group found "interesting indications" that the spirochete Noguchi showed them might be the cause of yellow fever, but decided to make no announcement. "Some tentative statement may be given to the public after Dr. Flexner's return. . . .If it should turn out that this is the cause of yellow fever the work of the Commission would become historic."[1]

Flexner returned in December and within the following few months he received three separate communications from Guayaquil transmitting various resolutions and commendations that had been authorized in recognition of Noguchi's work in Ecuador. In one communication, *El Telégrafo,* "representing the press of Ecuador, renders respectful homage to the illustrious Japanese, Dr. Hideyo Noguchi, discoverer of the microbic cause of yellow fever."[2] Flexner duly reported all of these communications to the executive committee, wrote a restrained acknowledgment to Guayaquil, but told the local press nothing.[3]

---

*Dr. Prudden was a member of the Board of Scientific Directors of the Institute from 1901 until 1924.

Noguchi's report to the foundation was guarded in its claims. He made only the most careful observations that the disease with which he had worked seemed to fit the descriptions of yellow fever given by other workers, and that the disease bore striking resemblances to Weil's disease as it occurred in Japan.[4] Yet he had already gone far beyond this labored conservatism in a brief communication published in the *Journal of the American Medical Association* in January in which he described the disease that had appeared in guinea pigs after their injection with blood drawn from cases reported as yellow fever. Noguchi said that positive results had been obtained from six of twenty-seven cases from which he taken blood, but that he had recovered a *Leptospira* that closely resembled that of Weil's disease and had grown that organism in pure culture "by special methods." Elaborating on the work itself and the importance of his special methods, Noguchi avoided any question of diagnosis.[5] In effect, he laid the basis for the whole ensuing controversy over his findings, even initiating what was to become his chief defensive weapon — his claim that he had used "special methods" — a claim that has enraged his critics ever since.

At the end of April, Noguchi reminded Flexner of his suggestion that Dr. Wenceslao Pareja, director of the hospital for infectious diseases in Guayaquil in which Noguchi had carried out his work, be invited to the institute for training to undertake therapeutic work with the immune serum that Noguchi was making.[6] Flexner arranged for Dr. Pareja's visit through the International Health Board of the foundation.

On June 1, three lengthy papers by Noguchi appeared in the *Journal of Experimental Medicine,* with the promise of more to follow. Although he presented an elaboration of the procedures he had followed and of the clinical and pathologic features of the disease, he really added nothing to the note that had been published in the *AMA Journal.* "That the present organism is closely allied to but immunologically distinct from that [Japanese] species of infectious jaundice has since been established and the experiments bearing on this point will be discussed in detail in subsequent papers."[7]

Noguchi had sent enough vaccine for the inoculation of 1,250 persons to Ecuador for use among the nonimmune population and foreigners.[8] Learning of this, Lebredo feared that a vaccine would come to be relied upon as a primary preventive measure before its value had been adequately established, and he expressed these concerns at the annual meeting of the Society for Tropical Medicine in Atlantic City. Half of the sessions of this meeting were devoted to yellow fever, and much of the time was taken up with presentations of the work of the Guayaquil expedition.[9] In a formal paper, Lebredo stated that he had been a member of that expedition, but tactfully said nothing about the physical conditions or medical facilities he had seen there. He reviewed the essential characteristics of yellow fever that had been established by the Cuban work in 1900: that the disease was transmitted only by a mosquito; that the blood of yellow fever patients was infectious only during the first three days after the onset of the disease; and that such animals as the guinea pig had been found to be not susceptible to the disease. Lebredo stated that Noguchi had not established the fact that the cases with which he had worked met these criteria, and until this fact had been established he believed that the results should be received with great caution. He emphasized his concern that the mosquito eradication program, which had been so successful in eliminating yellow fever from Havana and from the Canal Zone, would be abandoned in favor of a program of vaccination based on the organism that Noguchi had identified as the cause of yellow fever.*

Lebredo added the observation that the clinical picture that Noguchi had given of yellow fever followed almost word for word the description that the Japanese workers (Inada and his co-workers) had given of Weil's disease. With expressions of great respect for Noguchi, Lebredo nevertheless urged that much more evidence be assembled before Noguchi's *Leptospira* be accepted as the causal agent of yellow fever.[10]

Flexner was to write later: "Considerable hostility has always been shown by the Cuban Department of Health to the *Leptospira icteroides* view of the origin of yellow fever. The circumstances are unfortunate and personal, but unavoidable."[11] Miss Tilden says that the Cubans were annoyed that they had not been given more important positions on the investigating commissions.[12] For whatever reason, Lebredo's warnings were given little heed.

During the summer, the Rockefeller Foundation appointed a new group to undertake an expedition to Central America to determine whether yellow fever was being correctly diagnosed there. General Gorgas took charge of this expedition, with Dr. Pareja going along as a clinical observer. Although Gorgas had been in charge of the mosquito eradication programs of Cuba and Panama, he was not a clinician; Pareja, who had advised Noguchi in Ecuador and had spent the summer in training to use Noguchi's immune serums and vaccines, was hardly the best choice to view Lebredo's warnings with scientific detachment. On the contrary, "the party took with them about two litres of immune horse serum and enough killed culture for vaccination of 500 persons."[13]

By the middle of November, Noguchi had submitted eleven papers of his series on the "Etiology of Yellow Fever" to the *Journal of Experimental Medicine*. Paper number X of this series was devoted to a detailed presen-

---

* The term vaccination was used with much less precision during this period than it is today. Similarly, the word serum was often used when it was intended to refer to a specific immune serum or antiserum.

tation of Noguchi's comparative studies of the immunologic reactions of his newly discovered *L. icteroides* with *L. icterohemorrhagiae,* the organism causing Weil's disease.[14] His findings were inconsistent, yet he was satisfied to explain the inconsistencies by his belief that there was a wide variation in the susceptibility of guinea pigs to the disease and on the variations in the virulence of the organisms of both Weil's disease and the supposed yellow fever. Both of these arguments were to be found wanting by experiments of later workers, experiments that Noguchi could have performed at this time if he could have recognized the scientific necessity for doing so.[15] But Flexner must also have been satisfied; otherwise the papers would never have been accepted for publication.

While all this activity was going on, a yellow fever epidemic broke out in Yucatán, and Noguchi received an invitation from the Mexican government to come and investigate. "My trip was to study the Mexican yellow fever and see if it is the same as that existed in Ecuador," he wrote Madsen, adding, "besides, it was to test the value of immune serum."[16] Lebredo had gone to Yucatán as soon as the epidemic appeared and, as Noguchi must have learned fairly quickly, had found no *Leptospira.* Noguchi was anxious to go, but Flexner hesitated because of the disturbed political situation in Mexico at the time. Dr. I. J. Kligler, a young postdoctoral fellow at the institute, was to accompany Noguchi; the trip was planned and canceled several times. As late as November 26, Flexner wrote Noguchi (from his office down the hall): "At the present time I do not think it is either possible or advisable for you or Dr. Kligler to start for Mexico."[17] Within twelve days, however, Kligler was on the high seas bound for Yucatán, writing of the well-being of his guinea pigs and of his own pleasant voyage.[18]

Noguchi left for Mexico a week later. He planned to stop at Havana to talk with Lebredo and his colleagues who were the most outspoken critics of his work in Ecuador. There are hints that Flexner really did not trust Kligler and he may have conveyed some of his mistrust to Noguchi. If Noguchi was already uneasy at the prospect of his discussions in Havana, neither he nor Flexner would have wanted Kligler to hear criticism or be exposed to doubts. Whatever the reason for Kligler's being sent on ahead, Flexner wrote:

> Dear Dr. Kligler:
>   I am sending this letter by Dr. Noguchi in order that you may clearly understand that you are under his general orders. If he wishes you to come back with him you are to do so. If he wishes you to stay when he leaves, you are also to do so. In any case, you are to be precisely guided and instructed by him as to the work you are to do, how you are to do it, and when you are to come back. . . .[19]

The day after Noguchi's arrival at Mérida, Kligler wrote to Flexner:

Dr. Noguchi has kindly delivered your note. You will pardon me if I confess that I was a little peeved. I thought that you would credit me with the good sense to understand my relation to Dr. Noguchi. I appreciate too well the opportunity of this trip and the privilege of being associated with Dr. Noguchi to have acted in any way other than that indicated in your note. . . .[20]

On receiving this reply, Flexner still could not drop the matter, but his elaboration gives some indication of Noguchi's role:

[The letter] was written at the last moment, just as Dr. Noguchi was leaving for the ship, and was not intended at all to disturb you but to make clear a point which I had not emphasized sufficiently, namely that if Dr. Noguchi thought it was unnecessary for you to remain after he concluded to go back, that you were to submit to his wishes. . .since Dr. Noguchi felt some hesitation himself in leaving the question to decide itself and felt more comfortable in being clothed with authority, I dictated the note which he gave you. . . .[21]

Flexner wrote to Noguchi that he hoped that Noguchi was "patting Kligler on the back" since Kligler seemed a little upset.[22] Flexner continued to belittle Kligler throughout the next six months during which Kligler worked under the most trying conditions; indeed he was lucky not to have lost his life. He was responsible for whatever success the missions of these months achieved, which Noguchi acknowledged before the missions were over.

Kligler was the son of Jewish immigrants from Poland and had worked his way through college and medical school in New York. He was a tiny man, less than five feet tall, very admiring of Noguchi and, at this time, modest and naive. As Kligler's later career demonstrated, and as his son described him, he was both ambitious and idealistic.[23]

Noguchi's visit in Havana seems to have been less difficult than he had anticipated. Lebredo had made little effort to culture an organism from the blood of yellow fever cases in Mérida; since he had not been able to transmit the disease to guinea pigs, he considered it useless to pursue the organism.[24] Noguchi wrote Flexner, however: "It turned out that Dr. Lebredo had an unsuitable dark field microscope and we could not even see mouth spirochaetes when he tried to demonstrate. He has no oil immersion for the dark field work. . . .Dr. Guiteras now knows that the microscopic and cultural sides of Dr. Lebredo's work were faulty but he maintains that a negative animal transmission is difficult to understand."[25] Noguchi seems to have given no serious consideration to Lebredo's negative findings, but rather felt that he had put Lebredo on the defensive. He admitted that he had been treated with courtesy, and he felt that he had made helpful suggestions.

When Kligler arrived in Mérida, the yellow fever epidemic was prac-

tically over and he feared that he would have no material. He managed to
get some blood from a patient who died right after his arrival, and from
another with a mild infection. He made cultures and injected the material
into guinea pigs. After Noguchi's arrival a week or so later, one of the
guinea pigs fell ill and a spirochete was recovered. Noguchi identified it a
L. icteroides and wrote both Flexner and Dr. Heiser at the foundation in
detail of Kligler's role in this allegedly successful outcome.[26]

Meanwhile, Gorgas, Pareja, and others were appearing all over Central
America checking on reports of yellow fever cases, and Flexner was trying
to supervise Noguchi and Kligler from New York. Flexner had made a
substantial commitment to Noguchi's yellow fever findings, but he now
had little control over the jubilant Noguchi-Kligler team who responded to
calls from Gorgas as freely as they did to directives from Flexner. Flexner
asked that he be kept informed of their plans by cable, "notwithstanding
the expense,"[27] but their cabled reports could have brought him little
reassurance. He learned that Noguchi wished to have Kligler join him on a
trip to Mexico City, and that the possibility of Kligler's going on to Peru
had arisen, and he received a long list of the equipment, which included
three hundred guinea pigs, that Kligler would require for his work in Peru.
Decisions were made in Mexico City, Kligler started for Panama to accompany Gorgas to Peru, and Noguchi returned to New York.

En route to Panama, Kligler wrote to Flexner: "I presume I am to continue the yellow fever work, but I should like to have more detailed instructions from you. Shall I consider myself free to plan the laboratory work?
What am I do do if there are no cases?"[28] Flexner's ill-tempered reply
betrayed his own feeling of helplessness: "You say you presume that you are
to continue yellow fever work. Of course that is what you are going to Peru
to do. It is the only thing, as a matter of fact, that you are going to do." He
had to give Kligler some freedom to use his own judgment, but he garnished
it with all the restrictions possible from such a distance. ". . .you are really
under General Gorgas, and you will therefore wish to do precisely what he
thinks advisable and desirable. . . ."[29]

Flexner, so often referred to as the benign dictator who had to know and
be in control of every detail around the institute, had taken a dangerous
step in allowing these young investigators to take the field on their own, a
step that brought erosions of his control just when his need for that control
appeared most urgent. First, there was talk of further expeditions to Africa
under Gorgas just as soon as he finished in South America. Second, Alexis
Carrel, as a Nobel Prize Laureate, had proposed Noguchi for the Nobel
Prize, at least in part on the basis of Noguchi's work on yellow fever.[30] And
third, the Rockefellers had discovered Ivy Lee, who was to create the
modern concept of public relations with the Rockefellers as his first important client, and Flexner was working with Lee on the first announcement

to the general public of the foundation's work in yellow fever. Lee reported progress in a letter to John D. Rockefeller, Jr., which accompanied a copy of the article that was to appear shortly in the *Review of Reviews*.[31]

The article, which had been written by Dr. Prudden, but was signed only with his initials, dealt with yellow fever. The article began with the announcement that Noguchi and Kligler, both of the scientific staff of the institute, had gone to Yucatán to study the disease there. It described Noguchi's previous mission to Guayaquil, and recounted the history of yellow fever and the various attempts to deal with it in the past. Giving some details of Noguchi's background and "masterly command of cultural technique," the article stated that he was well acquainted with Weil's disease, which resembled yellow fever. "This had been first cultivated in Japan by the use of a special procedure devised by Dr. Noguchi." The article went on to explain the organism and bacteriologic techniques. Subtly catering to the intellectual reader's appetite for information on such esoteric matters, it carefully avoided the claim that Noguchi had discovered the cause of yellow fever, but left the reader to draw his own enlightened conclusions.[32]

Kligler had been sent to Peru because of an epidemic of yellow fever raging in the northern province of Piura. At the same time, bubonic plague had broken out in such massive proportions that the Peruvian government had recruited Dr. Henry Hanson from the U.S. Public Health Service in Panama to take charge of a control program. Hanson concentrated on the eradication of rats, the source of the bubonic plague, since that appeared to be the more devastating of the two problems. Hanson found the whole province suffering from both diseases, but the focus and highest incidence of illness, and the most deplorable sanitary conditions, lay in the port village of Paita, in the province of Piura. Hanson made his headquarters at Paita where he determined, after two months of desperate effort from January to March, that most of the poorer houses could be not be made rat-proof and would have to be burned. With the support of General Gorgas, who visited twice during this period, and of Dr. Henry R. Carter, assistant U.S. Surgeon General, Hanson persuaded the Peruvian national authorities at Lima to grant him permission to burn much of the village.[33]

Kligler arrived in Lima on March 3 and wrote that cases of yellow fever were still being reported from several villages in the province of Piura, three days by boat from Lima. He planned to prepare and sterilize all his equipment in Lima and leave for Piura on the first boat. He remarked in a letter to Flexner almost as a side issue: "You will no doubt be interested in the work done by Dr. Gastiaburu.* He went to Piura last September, studied two cases of fever and succeeded in isolating leptospira from one of

---

*Director of the Laboratory of the Health Department at Lima.

them. Then he noticed that a pig inoculated sometime ago and which
never showed any symptoms contained the Lept. in the blood. He,
therefore, examined eleven normal noninoculated guinea pigs and observed a Lept. in eight of them!" Kligler did not know how to interpret this information and seems not even to have surmised that the patient might have
Weil's disease rather than yellow fever or that the guinea pigs, also, might
have carried Weil's disease. Kligler wrote Flexner: "I don't think there is
anything in this, tho' it is possible that he did isolate the Lept. from the
yellow fever patient, his finding of a similar organism in normal pigs from
Lima makes that doubtful. I suppose that in order to dispel the doubt it
will be necessary to examine some of his guinea pigs. Do you not think so? I
am planning on my return from Piura to go over Dr. G's work with him
and at the same time examine some rats caught in Lima. What is your advice?"[34]

Flexner's cabled advice to Kligler was brief and to the point: "Confine
studies to yellow fever cases. Disregard spiral organisms in guinea pigs."[35]
In the presence of the many questions raised by Lebredo, and the massive
infestation by rats which by then had been identified as the carriers of
Weil's disease in Japan, Europe, and New York, such advice can now be
recognized as an emotional reaction on Flexner's part rather than a product of his scientific judgment.

Kligler arrived in Paita on March 12. He found Hanson struggling
against yellow fever as well as bubonic plague, and in spite of the conditions he decided to remain:

> The laboratory is a crude affair. There is literally no water in town.
> Whatever little there is is drawn from a river up country and distributed
> in barrels on burros. It is muddy and probably grossly polluted. . . .
> There is electricity at night but the current is alternating and so weak
> that thus far I have been unable to use my dark field. . . .The guinea
> pigs have fared very badly, but for the present I have a sufficient
> number for work here. . . .I have gotten blood from four typical cases,
> all in the second or third day of illness, made cultures and inoculated
> guinea pigs. . . .Paita is one of the principal ports of Peru. It has a
> population of about 3,000. It is merely a cluster of primitive huts with
> narrow corked alleyways. . . .The sanitary conditions are extremely
> bad. . . .I am not afraid of yellow fever, but do fear plague.[36]

Two weeks later Kligler wrote Flexner that the work had gone badly and
that none of the guinea pigs had developed a typical picture although he
had inoculated about seventy of them. Many of the pigs had died of
unidentified ailments.

> It sounds as though I am trying to find an excuse. . . .I do think I am
> guilty of bad judgement. . . .[I underestimated] the bad conditions and
> the delicacy of the work. . .I still have some of my cultures. I also hope
> to get a few new cases. It takes six days to go to Lima on account of the

quarantine, but the cultures ought to keep and remain infectious. . . .
Hoping you will not judge me too severely. . .[37]

Kligler had cabled the substance of this message, but even before his letter arrived, Flexner had decided to send Noguchi to rescue Kligler and the
work from failure. En route to Peru, Noguchi again stopped at Havana,
but this time without trepidation. He wrote that he demonstrated the
organism that he had found in Mérida and "They were overjoyed to see the
organism for the first time: the culture was very rich and showed highly
motile leptospiras." Noguchi adapted the Cubans' microscope for dark
field work by putting together pieces from various microscopes in the
laboratories and he gave them some cultures and uninoculated media.
Although the Cubans planned to start their investigations over again,
Noguchi felt that the laboratories were now being poorly supported and
that Lebredo was working under great handicaps.[38]

Kligler, meanwhile, was taking corrective action. He decided that part
of his failure at Paita was due to the deterioration of the rabbit serum that
he had brought with him for culturing organisms, and he now obtained a
rabbit and prepared a fresh batch of serum.[39] He also transferred his
laboratory to the hospital in the town of Piura, some fifty miles inland
where the dryer, more desert-like climate might be healthier for his guinea
pigs as well as for his cultures. In response to reports of yellow fever cases at
Morropon, a village of 1,000 people, about 65 miles further inland, he
traveled "12-13 hours on horseback; a killing trip" to Morropon where he
took blood samples from six supposed cases of yellow fever.[40]

At Morropon Kligler fell ill with what he assumed was yellow fever and
gave himself an injection of antiicteroides serum. His rapid recovery convinced him that his illness had indeed been yellow fever that had been
aborted by the serum, and six days after the onset of his illness he got back
on his horse and made that "killing trip" again, back to Piura. The second
day later, May 5, found Kligler waiting on the dock at Paita to welcome
Noguchi when his ship arrived.[41]

By the time of Noguchi's arrival, the rats had been burned out of their
hiding places in Paita and there had been no case of yellow fever or plague
there since April 16. Noguchi wrote respectfully of Kligler's trip to Morropon, and was cheered by Kligler's apparent success in using the antiicteroides serum to abort his own illness. Noguchi and Kligler went
directly from the ship at Paita to Piura by train and were suitably greeted
at the station by all the proper officials. They attended a reception in
Noguchi's honor that evening, and Noguchi wrote Flexner: "In the meanwhile all my things, including 7 boxes of laboratory supplies and 2 crates of
300 guinea pigs and many bags of oats were immediately attended and sent
to the Hospital where Doctor Kligler had established a laboratory. I went
there in the evening with the whole group of reception committee.

*Unidentified person, Hideyo Noguchi, and I. J. Kligler, Peru, 1920. Pro-
perty of Dr. David Kligler*

Everything was lovely there. The animals are kept in a nice room."[42]

The next day Noguchi and Kligler injected thirty guinea pigs with the cultures Kligler had obtained at Morropon, and Noguchi reported that although Kliger had worked for four weeks in Paita, none of that work was of any use. "Nobody can blame him. In fact, he did very admirably under the circumstances."[43]

During the next few weeks, Noguchi and Kligler examined specimens of convalescent serums from Paita and found them uniformly positive. Then the cultures made by Kligler at Morropon were examined and again Noguchi found leptospira where Kligler could not. Once again, Noguchi's experience, persistence, and skill in this kind of work led to positive results where others had met with failure. But once again, both men were acting on the assumption that the material came from cases correctly diagnosed as yellow fever, and Kligler was only too happy to be convinced that the frightful risks he had taken had not been in vain.

Rather than expecting any thanks for his efforts, Kligler continued to be admiring and grateful. He confessed to Flexner, "I have been wanting to write to you ever since I received your kind note of March 18, but hesitated. When I got your note I was too despondent to write; I felt quite sure that your attitude toward me had changed."[44] Such thoughts reflected the pressure Kligler had felt, partly because of his future career, just as others who worked at the institute at that time have described their feelings.

But for Noguchi things were altogether different. Planning to start for home, he enclosed a cable in a letter to Flexner, speaking of it as though it were a great mystery.[45] He had received the cable on his arrival at Paita and had answered by saying he must consult Flexner about its contents. It was an invitation to join an investigation of hoof-and-mouth disease in Scotland. Noguchi was flattered, and he wrote to Kobayashi in Japan to say that he was accepting this great honor.[46] Although Flexner later vetoed the trip, at this moment Noguchi was heartened.

A specially printed menu for what appears to have been an elegant farewell dinner in Noguchi's honor provides a strange contrast to the descriptions of Paita given by both Hanson and Kligler. How was it possible to provide such a formal and sumptuous repast in such a primitive village where even the water was brought in on the backs of burros and where the poorer houses had so recently been burned as the only way of getting rid of the rats? Kligler left no comment, but he did keep a picture of the assembled diners that shows that he had been invited, and he kept a copy of that menu.

While they were in Lima en route for home, Noguchi was approached regarding a position as director of an Institute of Hygiene that the Peruvians proposed to establish and equip to his specifications. Noguchi was of-

dr Kligler

ALMUERZO DE DESPEDIDA

Al Sr. Dr. Hideyo Noguchi

OFRECIDO

por el señor Prefecto
Del Departamento de Piura

Menú

Desert Soati
___

Creme de Valoy
___

Sup de Mer Maitre D'hotel
___

Petit Mignonet Gran Duck
___

Esperges Sauce Mousoline
___

Poulet Roti Treute
___

Ensalada de Saison
___

Fruit de Temp
___

De mi las Café

Koctail

Graves-Sant-Emilion
Champagne

PLOUS-CAFE
___

CIGARRETS.

Paita, Mayo 26 de 1920

4111—Imp. "La Mercantil"—Paita.

Menu printed for banquet in honor of Noguchi, Paita, Peru, 1920. Property of Dr. David Kligler fered a salary of $20,000 for this position.[47] No one else seems to have received the kind of offers and awards that came to Noguchi, but the impression he made on Hanson (in contrast to the tone of his letters to Flexner) may have been the image he presented to others as well. Hanson later wrote:

> We had all enjoyed the association with Noguchi and the little Japanese, with his charming eastern manners, had become quite popular with the doctors in the district. His success, the result of hard work and determination, appealed to the imagination of the men and more especially as Noguchi had overcome the handicap of a congenital malformation of one hand. . . .Forty-four years old at the time, he was at the height of his career. There was a certain reticence about him and whether from years spent in scientific research, where an open mind is necessary in order to observe new phenomena, or from some innate self distrust, he seldom made a statement which even verged on the dogmatic; there always crept into his conversation the conditional clause.[48]

Hanson clearly misinterpreted Noguchi's Japanese use of the conditional clause. Even though he spoke of Noguchi's eastern manners, he did not know that Japanese custom requires the courteous "may" when the speaker may really mean "positively is." Just as Hanson was delighted and misled by this linguistic courtesy that no Japanese could bring himself to abandon, so may other Westerners have found this a reflection of Noguchi's oft-proclaimed modesty and been similarly charmed by it. This was the other side of the Noguchi who had "proved beyond all doubt."

Flexner again was away when Noguchi and Kligler arrived back at the institute in New York, but he left them each a letter quite different from the other. To Kligler he had written: "You evidently worked under baffling conditions at Piura. But all is now well. The important thing was to avoid a flip-up—and that has been done. Now about your future plans. . ." Flexner explained that it was the policy of the institute not to keep the younger men on indefinitely, and that he had already recommended Kligler for some positions, among them one in Palestine.[49] To Noguchi he wrote: "I am very happy over the results you have achieved in Peru." He urged that Noguchi make no more expeditions during the summer and said that he would be disappointed and "really displeased" if Noguchi did not go up to the mountains for at least two months' rest. "You have had a hard year and a wonderfully successful one, and are entitled to a good holiday; and aside from that, for health reasons it is imperative." Noguchi was not to go to Scotland.[50]

On the day after his arrival at the institute from Peru, Noguchi wrote a long letter to Flexner. A request had come from Dr. Heiser at the foundation for serum and vaccine to send to Peru, and he told Flexner that he had an appointment to see Heiser the following day. "Dr. Heiser or you yourself may send as much serum or vaccine that may be desired by the authorities there. I shall have them ready at all times from now on."[51] But another problem took more explanation.

Among the areas in which yellow fever was prevalent in West Africa were the Gold Coast and Nigeria, both in the domain of the British Colonial Service. In anticipation of future work in Africa, Dr. Adrian Stokes came from England to work at the institute for a few weeks; he had already shown his interest in Weil's disease by collecting the organisms that had been sent from France during the war, and had done some cultivation work in confirmation of Noguchi's reports. Stokes returned to England while Noguchi was in Peru, taking with him some *L. icteroides* cultures and immune serum of both that organism and the organism of Weil's disease for further study. When Noguchi came into his laboratory, he learned from his technician that "Dr. Stokes. . .was somewhat confused with the behaviors of the antiicteroides immune horse serum (also those of anti-hermorrhagic serum) toward both species of leptospira." Noguchi wrote five pages of explanation to Flexner about the sources of Stokes's confusion.[52]

The serum problems, as Noguchi saw them, were twofold: first, there were atypical reactions because of what he called a rabbit precipitin in the horse serum; and second, the antiicteroides serum, which had been made for therapeutic purposes only, also contained some antihemorrhagic serum as well. Noguchi's explanation must have seemed reasonable to him, but it reads today as just frightening. It is full of such qualifications as "masking of the true reactions" by what he called deceptive phenomena. He spoke of modifications he had made to compensate for practical problems, for accidents that had occurred, and for errors resulting from misunderstanding of his instructions — modifications that "do not make any difference" and hence could be disregarded. He said it did no harm to have a little antihemorrhagic serum mixed in anyway since there were occasional cases of Weil's disease in South America, and he emphasized that he had made it clear that these immune serums were not suitable for differentiation of these organisms.

It was bad enough at the time to be manufacturing immune serum and vaccine with all these ingredients, added intuitively and recorded only in Noguchi's head; in the light of more recent knowledge — that *L. icteroides* was identical with *L. icterohemorrhagiae* — there can be only horror that such explanations would be offered for the inconsistent performance of the immune serums. These arguments ended with "Do you not think the addition of the icterohemorrhagiae was just as justifiable as any other combinations of more than one antitoxin or antibody? At any rate, the serums are principally anti-icteroides."[53]

There is no indication of Flexner's appraisal of this argument, but after he returned from Europe and settled down in New Hampshire for the summer, Noguchi came from Shandaken to visit him. After Noguchi's return he wrote Flexner to answer several questions that had been discussed and gave an accounting of the amounts of immune serum that he had available for bottling, of vaccines that had already been shipped, and of arrangements for future production.[54] Flexner must have been satisfied with the procedures being followed, but he had long since forgotten his orders that Noguchi was to take a complete vacation.

Noguchi was expecting a visit by Mrs. Flexner and the two Flexner sons on August 18, and he was disconcerted when Dr. Flexner appeared with his family. (A member of the institute says that Noguchi always became agitated when Flexner appeared unexpectedly in his laboratory, dropping test tubes and knocking over bottles that he never did otherwise.) The day after this visit, Noguchi dashed back to New York in response to a report of new problems with the vaccines, a report that probably had reached Flexner's ears and had prompted his visit. Miss Tilden had returned from vacation to find the laboratory boys making up vaccines and had sounded an alarm that resulted in Heiser's sending a cable to the foundation representative in Guatemala, "Noguchi withdraws Guatemala vaccine arriving August 24, serum all right."[55] Miss Tilden reports that this was the only time in her years of association with Noguchi that he lost his temper with her, but she claims that he was angry because she had not reported the matter to him earlier. At any rate, she kept out of Noguchi's way for a few days, and he later apologized for the manner in which he had spoken to her about it.[56]

After returning to Shandaken, Noguchi wrote a report of the episode to Flexner, giving Flexner his personal assurance that the immune serum that had been shipped was all right. Noguchi also fired one of the laboratory boys, berated the other, and solved the problems created by the rapidly increasing demand for vaccine by staying and doing the work himself.[57] He was annoyed at the manner in which the problems had been brought to Flexner's attention, but asked that Miss Tilden's services be assigned to him on a full-time basis.[58]

Before winding up his frequently interrupted vacation at Shandaken, Noguchi also wrote Flexner in response to an inquiry concerning the advisability of sending Kligler to Vera Cruz to help out with an epidemic of yellow fever there. Noguchi spoke well of Kligler, yet said Kligler should not be chosen for this assignment:

The only reproach he may deserve is that he cabled his work too soon, before he did all he should have done. In fact, he was entirely premature in announcing his Payta work as negative....He also succeeded in isolating the icteroides in Mérida. Therefore, from the standpoint of his ability, under favorable conditions, to isolate the organisms there is no question in my mind. Only drawback, if we may call it so, is that he breaks down more easily than I when facing preliminary failures. I do not know how good the laboratory facilities are in Vera Cruz, but there will always be plenty of difficulties to overcome when starting this kind of work in a less convenient locality. I am inclined to think that it may not be altogether a certainty that he shall succeed.[59]

So Kligler went to Palestine. Yet, when Flexner made his annual Scientific Report to the Corporation of the Institute in the fall of 1920, he gave extensive and detailed attention to the work in foreign fields in general, and the Noguchi-Kligler work in particular. He told of the work they had done in both Yucatán and Peru, with a lurid account of Kligler's trip to Morropon, returning with the cultures just as Noguchi arrived, thus "the final studies which proved positive and conclusive were carried out. This success, snatched as it was from imminent failure, which would have led to an attitude of skepticism throughout South America, perhaps throughout the world,. . . produced an almost dramatic effect in the opposite direction." Citing immunologic work that Noguchi had done in the meanwhile, Flexner concluded "The etiologic relationship, therefore, of the *Leptospira icteroides* to yellow fever has been placed by these new studies on a very substantial basis."[60]

Flexner served as president of the American Association for the Advancement of Science during 1920. When he took office in December 1919, he gave an address that was not published in the proceedings but was reported by the *St. Louis Post-Dispatch*. In this address Flexner spoke of the problems with infectious diseases, of the efforts of the Rockefeller Foundation to eradicate yellow fever, and of Noguchi's finding the *Leptospira*. "It now seems possible that yellow fever may be the first disease to be eradicated by science."[61]

Flexner's address at the end of his tenure as president, however, dropped all pretense of conservatism. He chose to review the progress of bacteriology during the previous twenty-five years, and he developed the subject to the climax of his own work with poliomyelitis and Noguchi's alleged contribution to that effort as well as Noguchi's own discovery of the yellow fever organism. Flexner discussed the nature of viruses, much as he had done in Japan, but gave credit to the great improvements in the microscope through the new technical devices for dark field illumination. Claiming that Noguchi had made major advances in the knowledge of viral organisms, he stated that Noguchi had demonstrated that both the yellow fever and poliomyelitis viruses were, in fact, only smaller, more difficult to visualize, forms of previously known organisms; that yellow fever was only a tiny spirochete, and poliomyelitis could be cultivated to form visible clumps when subjected to the same cultivation techniques as Noguchi had developed for spirochetes.[62]

Now there was no turning back for either of them. Flexner had committed his own and Noguchi's scientific integrity and judgment, not just to a report of experimental findings but to the interpretation that their work justified a revised concept of the nature of viral organisms.

18

The Institute in the Twenties

WHEN DR. FLENXER TOLD THE MEMBERS OF THE BOARD OF CONTROL about Noguchi's brilliant results in his expeditions to Yucatán and Peru, he reported also on other expeditions to Africa to study sleeping sickness, on ambitious efforts to find a therapeutic agent for syphilis, on a program that he hoped would lead to a therapeutic serum for pneumonia, and on plans for a new laboratory in which Carrel was going to expand his work with tissue cultures. Flexner said nothing of the basic research that was in progress in the laboratories of Jacques Loeb, Phoebus Levene, Oswald Avery, or even the newly arrived Landsteiner from Austria who had transmitted poliomyelitis to animals before Flexner ever got started in this work.[1] Flexner seemed to believe that the institute had gone through its period of trial, and that all the flashy activities of its superstars would bear the fruit that was the original objective. He seemed quite unaware that anyone would think differently.

In October 1914 a formal dinner in honor of Dr. Flexner had celebrated the tenth aniversary of the institute, and in response to the flowery speeches, Flexner said "The first years were nervous ones for all concerned in the actual work and I suspect the Scientific Directors did not escape this feeling of uncertainty which may be expressed by the slang phrase, the necessity of 'making good.' "[2] Such occasional dinners, when formal speech-making was in order, along with equally formal teas, were the only social activities of the institute during Flexner's tenure as director. One member has described them as funereal; no alcoholic beverages could be served because of Mr. Rockefeller's rigid abstemiousness. Flexner's attempts to lighten his remarks with wit were pedantic failures. The diners were grateful for such relief as came from Noguchi's earnest attempt to handle the English language as he directed his remarks at the 1914 dinner to the spirit of friendship that flowed from Flexner's benign guidance. "The fine brotherly feeling has never left us. . . .There is such a good fellow-feeling among the workers, and I think all the men who serve under Dr. Flexner must have felt just so as I feel."[3]

Members recall the exhilaration of the younger men of those days when the dramatic Carrel appeared in the institute dining room for lunch wearing his specially designed surgical cap and gown — colored black to cut down the glare; the thrill when they were offered a custom-made cigarette from the gold case of Lecomte du Nouy; and the sense of living on a higher plane as they listened to the erudite discussion of the work in progress in the laboratories of these distinguished men of science. Yet they mention that a real atmosphere of informality prevailed only when some of the staff adjourned to the hospital after lunch for friendly chess games. In these, Noguchi participated with his own rapid-fire style of play. And it must be added that the free-floating wisdom of the lunchroom was confined to but one of three dining rooms. There was a second dining room for nonprofessionals, and a third for women.

A secretary who started work at the institute at the age of sixteen and who continued working for the institute and the foundation until her retirement, tells of Flexner asking her where she took her lunch and how much it cost. When she told him that she spent twenty cents a day in one of the staff dining rooms, he said this was too expensive and he would make other arrangements. She was to bring her lunch from home and eat it at her desk where he would have her served with a beverage and dessert without charge. Flexner's real concern, she thought, was that she was too young to be exposed to the worldly conversation of the dining rooms.[4]

Just as Noguchi told Flexner what he wanted to hear about the atmosphere of good fellowship, others may well have done likewise; but there is much to suggest that even when people tried to tell him otherwise, he did not listen. But there was one voice that Flexner could not avoid hearing, that of the troublesome young Paul de Kruif who arrived on the scene in September 1920. De Kruif undoubtedly had the ability to continue the successful career in science that he had started as a student of Novy at Michigan. Driven, however, by what he described as the mixed-up madness in his head to become a writer as well as a scientist, he was at once a fighting idealist as a scientist and a social muckraker as a writer. When Flexner heard what de Kruif, the writer, had to say, he fired him at once. Before that time, de Kruif did valuable work in the laboratory. He lasted two years, and his version of the experience gives an inkling of what others were saying privately, or just thinking, about the atmosphere of the institute at this time.

Professor Novy was probably the most able and productive American bacteriologist of the first quarter of this century. Flexner had offered him a position at the institute in 1907, but Novy refused it. Peyton Rous* had gone to work with Novy after finishing his graduate work at Johns Hopkins, but when Rous found that he could not stand the Michigan climate, Novy told him of openings at the Rockefeller Institute. Novy added his opinion that the place was supported by "dirty money" but left the decision of conscience about that aspect to the individual who would have to live with it.[5] Novy felt the same about it when de Kruif decided to go East because of his urge to come into association with writers, particularly H. L. Mencken. De Kruif adored Novy and later wrote:

> It was not so much that he would be sorry to lose me — my turbulence had at times been a source of embarrassment to him. It was hard to say I was leaving him for The Rockefeller because Dr. Novy took a dim view of Rockefellerian bacteriology. . . .Now what would happen to me in the streamlined elegance of The Rockefeller, where sensational scientific events were often portended but seldom came off? What would become of my discipline absorbed from the dedicated, austere Frederick Novy when it was exposed to the atmosphere of a scientific emporium where it didn't very much matter if one made mistakes what with Rockefeller prestige backed by the big money? What would happen to the scientific conscience of Novy's boy, Paul, when he consorted with The Rockefeller star, Dr. Hideyo Noguchi? Noguchi, a gay little Japanese, had earned fame resting partly on his discovery of a spurious spirochete that turned out, alas, to be definitely not the cause of yellow fever. You forgave Noguchi because he was a nice little guy, but Novy, the master of spirochetes, took Noguchi apart without mercy.[6]

De Kruif's ideal at the institute was Jacques Loeb. "He gave the institute a high scientific tone." In spite of his cynicism, de Kruif, too, was captivated by the spirit of the dining room. His mixed feelings revolved around the figure of Flexner.

> Among the eaters in this scientific beanery the most picturesque was the enigmatic Japanese, Hideyo Noguchi, a scientific Trilby, the creation and at the same time the slave of his Svengali, Dr. Simon Flexner. In the winter of 1920 I was asked to hitch my wagon to Noguchi's star. . . .Suddenly — it was a bolt out of the blue — Dr. Flexner asked me if I'd drop that research [in pneumonia] and go to Vera Cruz, Mexico, where a yellow fever epidemic was said to be raging. Purpose of mission: to isolate and bring back strains of *Leptospira icteroides*, claimed by Noguchi to be the cause of the dread Yellow Jack. Would I like to? I wouldn't like to. . . .I still believed Novy's view of Noguchi.[7]

---

*The Rous sarcoma was the first malignant tumor produced experimentally in animals in a laboratory.

De Kruif's career as a scientist bloomed and died in the two years he spent at the institute. His work with pneumonia in rabbits led to his discovery of a true bacterial mutation, and this promising beginning led de Kruif to work with his friend John Northrop on the chemical mechanism of bacterial agglutination. De Kruif was fired for reasons that had nothing to do with the quality of his work; it was inevitable that he and Flexner would tangle. Two versions of the process through which de Kruif was separated from the institute and scientific work to emerge as the first important science writer, complement each other in describing the relationship between the director and the rebellious young idealist. Both versons were written about forty years later: the first, by de Kruif himself in his memoirs, is still idealistic and aggressively irreverent; the other was written by Dr. Corner in his history of the institute, the Establishment view with a doubly capitalized E. They agree in their statements of the facts; each is a true statement as far as it goes. But neither version is complete without the other, and the point of their divergence comes in the personal responses of the writers to Flexner and the spirit of trust and friendship that he insisted could be created by fiat.

The facts are that de Kruif wrote, anonymously, the section on medicine for Stearns's *Civilization in the United States: An Inquiry by Thirty Americans*,[8] and he followed this with a series in *Century Magazine*.[9] All of these articles were indictments of medicine as it was often practiced. Corner wrote:

> . . .he illustrated what he considered the unscientific quality of much research then done by physicians by an example drawn from the hospital's studies — not identifiable as such by the general reader but sufficiently obvious to the Institute's people, some of whom knew de Kruif had written the article. Flexner, discussing staff problems a few years later in a confidential report, said that there had been one case in which a scientific worker overstepped the bounds of fellowship: "Reminded of the original understanding under which he accepted appointment, he at once proffered his resignation."[10]

De Kruif had known that his days at the institute would be over as soon as the anonymity of this *Century* articles was broken, and he even hints that he wanted it that way. Of the final confrontation with Flexner, de Kruif wrote:

> Then, with a mixture of relish and regret, the good doctor launched upon a discourse. I interrupted him, quickly and politely. "Please, Dr. Flexner, may I have stationery and a pen? To write you my resignation from The Rockefeller Institute, to take effect today, September 1, 1922? What's done is done and I'm sorry to have upset you." He gave me pen and paper and the resignation — our American euphemism for firing — was accomplished in a single sentence and I rose to go. "Please sit down, de Kruif, I'd like to talk to you," said Dr. Flexner. . . .Then I

listened in astonishment as he explained his indignation. Of course he was proud of the accomplishments of the Institute—including my own, he interpolated with an almost friendly smile—but the Rockefeller's distinguished achievements were not what had made him most happy throughout his years as its director. . . .No, it was the harmony, the serenity, the brotherhood with which the staff had always worked together. Like a big, happy family. My amazement was due to our deep difference in point of view as to what it took to make real science. My belief demanded that in science every man, all out honest, must call them as he sees them. . . .Who'd ever heard of artists or composers or writers who amounted to a tinker's damn working together under one roof in an atmosphere of brotherhood and loving kindness and live and let live?. . . .Dr. Flexner now became concerned about my personal fate. He would like to change the date of my resignation from September 1, 1922, to March 1, 1923—that would give me six months' pay while I looked around for a position. On Dr. Flexner's part this was pure kindness. . . .I stood up. "No, thank you, Dr. Flexner, let's make a clean break. The resignation stands, as of today."[11]

De Kruif went on to his own future as a popular writer of articles and books on various scientific themes, one of which became a popular play, *Yellow Jack*, based on the work with yellow fever of Walter Reed and his associates in Cuba. Most notably, de Kruif provided the material for Sinclair Lewis's novel *Arrowsmith*, which is frequently considered, and as frequently denied, to be based on personalities of the institute. De Kruif admitted that the character in the novel Terry Wickett, the young man with the great future yet with a sense of reality, was a composite of his friend and collaborator at the institute, John Northrop, and his assistant, T. J. LeBlanc.[12]

Like others whose work was eventually rewarded with a Nobel Prize (Rous, Landsteiner, Carrel), Northrop had the ability to isolate himself from human conflicts. Jacques Loeb, who may have been the most productive and consistently creative of them all, never did get the Nobel Prize although Flexner nominated him at least nine times.[13] Both Loeb and Landsteiner ran their laboratories much in the German style; Loeb had come to the United States in 1910 and had been at Byrn Mawr,* Chicago, and Berkeley before accepting an appointment to create a department of experimental biology at the institute in 1923. Although the young men working with Loeb gained a great deal as they carried out the sophisticated assignments he gave them, only two remained at the institute, one of whom was Northrop.

Landsteiner came to the institute in 1920 in the aftermath of the destruction from the war in Europe, and he continued work that was essentially an outgrowth of the work he had done all his life. Studies in blood

---

*Under the presidency of Mrs. Flexner's sister, M. Carey Thomas.

groups, the antigens of blood and immunological specificity of the proteins, led among other things to Landsteiner's important discovery of the Rh factor, which he and his associate Wiener named for the Rhesus monkey in which it was first found. Landsteiner is regularly described as a *Geheimrat* — one man claimed that Landsteiner designed his laboratory in such a way that he could watch what all his assistants were doing all the time.

In spite of the vaunted free spirit of trust in the discussions around the lunch table at the institute, everyone was aware of private feuds that went on all the time. Notable was that between Rous and James Murphy over proper acknowledgment of the work that has come to be known as the Rous sarcoma, in which Murphy and many of his supporters felt that Murphy played an unrecognized role. One man who left the institute saying that he could not work with Rous because Rous wanted everything his own way all the time returned only when arrangements were made for him to work with someone else. This individual said that in those days Rous found it difficult to get along with Jews; that Rous was "in his Aryan period at this time — not really anti-Semitic, just afraid."[14]

These reports are cited not as necessarily factual in themselves but as indicators of the undercurrents of hostility and anxiety that flowed beneath the surface of calm and open friendship which Flexner found so desirable. Such reports came from men who thought highly of the institute and remained loyal both to it and to Flexner. These men are unanimous in their belief that the institute owed its success entirely to the administrative skill and wisdom of Flexner. They described him as a dictator, some said benign, but they felt that such a project could succeed only under a dictatorship of this kind. One man said that Flexner would do anything — no matter how unscrupulous — if he felt it was in the best interests of the institute.[15]

Northrop emerged on his own after Loeb's death in 1925, and was always aggressive in working as a loner. According to de Kruif, Northrop believed that "the great days of the American continent were done when the Pilgrims landed and began to bother those poor Indians; everything after that was a downhill road to doom."[16] Northrop hated cities and persuaded Flexner to let him move his laboratory to the Department of Animal and Plant Pathology at Princeton in 1926; when this department was closed, Northrop refused to return to New York and transferred to Berkeley, where his laboratories continued to receive the support of institute funds. When Northrop was awarded the Nobel Prize in 1946, both the institute and Berkeley added him to their lists of Nobel laureates.[17]

But Northrop, too, was loyal to Flexner. In 1972, having retired to the Arizona desert, Northrop said that he "minded his own business" at the institute and did not know anything about anyone else; that he had always told Flexner exactly what he wanted and Flexner had always given it to him and otherwise left him alone "so we got along just fine." Northrop also said that de Kruif had always wanted to be a writer but would not comment on anything de Kruif had written. Of Noguchi, he said, "the little man made a mistake and that's all there was to it."[18]

Dr. Flexner's consistent nominations of Jacques Loeb for the Nobel Prize suggest that he recognized Loeb's work as the most likely to receive favorable consideration by the Caroline Institute directors. Whether he believed that Loeb's work was the most important of all that was going on at the institute or whether he thought it would have a stronger appeal in Sweden because of its fundamental nature, Flexner did not say. Flexner certainly made a distinction between the work he suggested for such recognition and that which he reported to the general scientific community at the AAAS meeting and to the governing board of the corporation.

In spite of Flexner's efforts to maintain an aura of scientific restraint in publicizing the work of the institute and the foundation, there was an increasingly manifest divergence of opinion, both in the scientific community and in the general public, about the integrity of his administration and the trustworthiness of his reports of scientific achievements. As the search for sensational discoveries that would have immediate news value became clouded by so many failures, the dirty money image of the institute took on more life. Stories circulated about the pressure on scientists to work in assigned areas that would bring the kind of results that Rockefeller wanted (at least one of these stories was told of Noguchi's work);[19] yet just as frequently the loyal members rose to deny them.

It was easy to say that outsiders were envious of the obviously lavish support that workers at the Rockefeller institute could command, but that defense could not carry the burden that was increasingly thrust upon it. Dr. Max Delbrück believes:

. . . the Rockefeller Institute's great period, during which it dominated medical research in the U. S., ended in the 20's. It ended because its planners had concentrated too much on infectious diseases and neglected nutrition. It failed to develop strength in biochemistry and completely lost out when vitamins and hormones were discovered. Only a little later sulfa drugs and antibiotics came in, further diminishing the importance of infectious diseases as problems of medical research. Avery's discovery in the 40's that the genetic transforming principle is DNA was perhaps the most momentous discovery made at the Rockefeller. It happened, as it were, in a little corner, just as a few years later the Double Helix was discovered in a most inconspicuous annex to the great Cavendish Physics Laboratory in Cambridge. All that this goes to show is that no institute, no matter its financial and scientific potential, can foresee the great surprises or predict the great new developments arising from them.[20]

In 1910, Abraham Flexner, Dr. Flexner's brother, completed a study of

medical education in the United States and Canada that he had undertaken for the Carnegie Foundation. This study condemned the level of
teaching in most of the medical schools.[21] The General Education Board
of the Rockefeller Foundation employed Abraham Flexner to implement
the recommendations of his report by offering matching funds to schools
that accepted its standards.[22] This was dramatically effective in raising the
quality of medical education in that preclinical studies were vastly improved,
full-time faculty positions were provided, and research laboratories
became an accepted part of most schools of medicine. Effective as these
changes proved to be, however, they depended on the judgment of human
beings. The errors in judgment that occurred inevitably influenced the
new approach to medical education, establishing attitudes and practices
that are still a burden on the profession today. Many of the appointments
to chairs of teaching and research were Rockefeller Institute men, and the
selection of some of them has been the subject of controversy. It can hardly
be said that everything that has proved unfortunate in medical education
stems from the selection of these men; but de Kruif's complaints about the
institute have been repeated many times in relation to American schools of
medicine.

The total effect of the fifty years of the institute is still a subject for
discussion, the leading spokesman for its positive contributions being,
naturally, Dr. Corner. A broader view of the matter was expressed by Dr.
Zinsser in writing of the development of medicine in the United States in
the first quarter of the century. Zinsser was a wordy man; yet his many
years of varied experience in private practice, in public health work, and
in teaching and research, make his words worth repeating even at length,
if only for the ambivalent attitude they reveal.

It was during these years, also, that medical research in America began
to blossom forth with the new vigor instilled into it by the golden showers
that were falling upon it in benefactions unprecedented in the history of
science. . . .There had been groups, like those at Johns Hopkins, a few at
Harvard, and the school of Victor Vaughn at Michigan. . .but resources
had been limited and laboratories few. . . .Research institutes, were
founded in Philadelphia, Chicago, and New York. The organization of
The Rockefeller Institute. . .had an especially powerful influence as it
continues to have today. . . .Opportunity was made available faster than
the brains needed to take advantage of it could be mobilized. There was,
at first, much rediscovery of what was already well known in Europe
and, in many places, halftrained people in magnificent laboratories
were sitting on sterile ideas like hens on boiled eggs. But there was much
earnest effort, great velocity of improvement, and by the time of the
World War, American medical schools and research institutes could
hold up their heads with any in the world. . . .And in this transitional
period we had the good luck to have leaders like Welch, Simon Flexner,
Christian Herter, and others, whose encouragement of talent and en-

thusiasm wherever they found it makes most of us of my generation deeply their debtors. . . . Of course, the research world is not devoid of a certain amount of hocus-pocus. There has been, and is, a good deal of personal and institutional rivalry, growing as public interest increases; and this leads to premature publication and to the frequent ballyhoo of relatively unimportant stuff as the work of genius. . . . Indeed the pressure for premature publication has ruined a considerable number of good men, who might have built up sound and permanent reputations. From some of our greatest institutions for research, within half a generation, have come utterly erroneous publications announcing such major discoveries as the cultivation of poliomyelitis virus, of the syphilis organism, of a bacillus of epidemic influenza, of a yellow fever leptospira — all of these widely publicized and applauded, few of them ever corrected from within. . . . We have probably passed through the worst of this. It was one of the growing pains of medical research.[23]

Strangely enough, although Zinsser named Rockefeller men and the institute itself as the country's most influential and gifted leaders, the examples he cited of erroneous pulbications are all to be laid at Flexner's door (three of the four leading through Flexner directly to Noguchi). Does Zinsser speak hypocritically? Why did he pick on this institute? Was it really different from others? Was Noguchi one of the good men who were ruined? Does Zinsser speak with the ambivalence that is with us yet today as American medical practice and research are touted as the best the world has ever known in the face of such contradictory facts as the level of care available to the average citizen — in the face of the growing popular resentment toward the profession?[24]

The science and practice of medicine has always been very much a part and reflection of any society in which it functions. As that society vacillates between its search for a workable morality and stability on the one hand and the opportunity for rapid economic and social expansion on the other, the issues between the ends and the means become clouded. Zinsser dealt with American medical research as though the Rockefeller Institute should have represented the best because it had the greatest financial support; and, although he considered the delinquencies of the institute as only growing pains, he seemingly expressed a greater disappointment as he acknowledged none of the truly constructive work that was accomplished there and pointed out transgressions in no other institutions.

In scientific and technical fields that require sophisticated training to judge the competence and even the integrity of professional colleagues, it is difficult to keep personal insecurities and foibles out of the judgments that are made. In such a demanding relationship as that between physician and patient, and by extension between the medical research worker and the public so eager for medical miracles, all the insecurities of human relationships become grossly exaggerated.

Noguchi, like many other scientists of his day and all the days since, was

*Visiting commission of Japanese medical scientists. Front row; left to right;
Dr. Sahachiro Hata, Dr. Hideyo Noguchi, Baron Yoshihiro Takaki, Dr.
Simon Flexner, Dr. Konnosuke Miura, Dr. Rufus Cole, Dr. Keinosuke
Miyairi, Dr. Akira Fujinama, Dr. Mataro Nagayo. Rear row, left to right:
Dr. Alexis Carrel, Dr. Karl Landsteiner, Dr. E. Vincent Cowdry, Dr. Jac-
ques Loeb, Dr. James B. Murphy, Dr. Peyton Rous, Dr. Alfred E. Cohn,
Dr. Wade H. Brown, Dr. Walter A. Jacobs. The Rockefeller University
Archives*

acutely aware of the value of citations by his professional colleagues for the attainment of eminence. The Nobel Prize, which was first awarded in his day (1901), soon became the greatest status symbol of all. It was a far larger monetary prize than any that had ever been established, which may have been the basis on which its prestige grew so rapidly. The relationship of American men of science to money, the identification of their status through the level of their salaries and research funds was, and still is, equally a reflection of everything else in the culture. As a scientific career became an increasingly well-recognized opportunity for upward mobility in society, interest was increasingly focused on the prestige and the financial rewards. The foibles and talents of the administrator who wielded decision-making power in financial matters, the effectiveness of the scientist himself in promoting his talents, and the institutional and personal rivalries of which Zinsser spoke — all took on dimensions far beyond the "hot" science so affectionately proclaimed by de Kruif.

Noguchi had joined American academia with an intimate and resentful knowledge of the way things were done in Japan, where positions were available only to those whose family connections made them eligible for acceptance into this privileged group. The Japanese perception of teaching and research as a means of earning a living, supporting a family, or improving one's position in society was almost nonexistent. Noguchi had tried to buck this system, both for the important privilege of being a part of a group and for the opportunity to make a living, but he had failed. In the United States, the opportunity to become a part of the "Rockefeller group" had even more profound meaning for him than for others. In his early days Noguchi had been in constant financial trouble, even after his earnings should have been ample. He was aware of his shortcomings in the handling of money well before he faced the shortcomings in his training. When he first spoke of the Nobel Prize, he wrote Kobayashi that it would be a financial help as well as a professional honor.[25]

It is to be expected that Flexner would run as complicated a show in administering the financial resources of the institute as he did with the spirit of friendship. By the 1920s the institute had grown to the point where Flexner felt the need to put its fiscal affairs on a more formally organized, businesslike basis. His concern for the tiniest detail may be seen in such letters as Noguchi received twice a year:

Dear Dr. Noguchi:
 In going over the stamp account for the half year from July 1 to December 31, I find that the following envelopes were used for your correspondence:

	Official		Personal
2¢	5¢	1¢	2¢
134	47	23	19

Mr. Spies will now charge these envelopes to your accounts in the usual way.[26]

Thereafter every January brought detailed complaints that Noguchi had once again exceeded his budget, initiating a correspondence that often did not resolve the problems until April.[27] But none of these problems produced anything like the verbiage that salary matters brought forth. In the period immediately following the war, it became apparent that an inflationary trend in the economy would rob the institute of some of its top staff members unless countermeasures were taken. On the chance that the inflation was only temporary, neither Flexner nor the board members were willing to commit themselves to a policy they might later regret. They devised a temporary salary supplement that became a permanent raise later for most staff members, but for Noguchi this raise continued in part as a supplement for the rest of his career.[28]

Several of Noguchi's colleagues have suggested that part of Noguchi's salary was paid him without the knowledge of his wife; that Mrs. Noguchi was considered unreliable because of her drinking. No one knows what reason Noguchi gave Flexner for his desire that the full truth of his earnings be kept from his wife. We have seen that Flexner had been concerned for years about Noguchi's improvidence. Probably we will never know the dimensions of Mrs. Noguchi's unreliability.

Dr. Kanematsu Sugiura, who visited the Noguchi home frequently during the twenties and who knew of Mrs. Noguchi's drinking, says that the apartment was well kept and orderly. He, too, was aware that Noguchi was concealing some of his earnings from his wife, but says that Noguchi wanted to have some money with which to indulge his own whims without consulting his wife.[29] Although Rockefeller Institute staff members firmly deny that Mrs. Noguchi was socially unacceptable to the wives of other staff members, none of the institute wives ever visited the Noguchi home in New York nor did she ever visit theirs. And only the Flexners ever visited Shandaken. When older members speak today of Mrs. Noguchi's having been a maid in a boardinghouse where Noguchi lived, their tones convey their contempt for the woman Noguchi married.

Noguchi and his wife were not the only ones who had problems in finding their way in this pretentious world. Flexner, himself, had risen from a humble background, and he benefited from the guidance of a sophisticated wife in avoiding the pitfalls within New York's code of social conduct. Flexner was concerned over the many occasions in which younger staff members had trouble. Many of them came from small towns with wives who were delighted to join what seemed to be a glamorous life. But that life often led them into social situations they were not equipped to handle, and the result was embarrassment, tears, and recriminations.[30]

One young man, unmarried at the time he came to the institute in 1926, describes the social climate in the fondest terms. He has tremendous gratitude to the institute for giving him the opportunity both to enhance his education and to gain an expanded view of a world of which he had never dreamed. He arrived with a dewy-eyed vision of creative men working in blissful harmony seeking the deeper truths, and he got a liberal education in prima donna behavior, competitiveness, tension, and backbiting—and he loved every minute of it. He has enormous respect for Flexner, both as a scientist and as an administrator, not at all contradicted by Flexner's giving one incompetent man a job to please an important friend of the institute. This man knew Evelyn Tilden well, and spoke of the almost overwhelming problems she faced in getting a Ph.D. and the respect to which she is entitled for achieving it. Columbia University accepted few graduate students of either sex, and was extremely reluctant to admit women. He hints that some people believed that Noguchi and Miss Tilden were lovers, but says that that sort of thing was none of his business. "These Rockefeller scientists worked hard and played hard."[31]

Now what can be said in summary? Everyone who entered the door of the institute was soon caught up in a pretty heady environment and manifested his own anxieties according to the dictates of the personality with which he came. Being a part of a self-styled elite, chosen to lead the world toward the betterment of mankind, identified with the responsibility of converting all that money into goodwill and prestige, and seduced into fantasies that all one's dreams might indeed become realities—how much can one human being stand? Those who could turn away and lock themselves in laboratories were, perhaps, the fortunate ones. To come out and look around was to court disaster.

Yet even those who could turn away had a commonality in culture and education that gave them support through their personal feuds, rivalries, and egocentricities. Noguchi was bound only to the institute, and bound there only by his personal ambitions and his special relationship with Flexner. As poorly prepared to exercise independent judgment as it was possible to be, Noguchi stayed in his laboratory as much as he could. He stayed there all day and all night much of the time. But he really did not know what to do when he was there, and he did not know how to learn. When he did come out, he observed the status game being played on a somewhat freer basis than he had known it in Japan. He learned some new rules of the game, and he played pretty well in defense of his position. But where other members of the institute found the nourishment for self-correction and growth, Noguchi was isolated—an ornament on the institutional body rather than a functional part of it.

19

Noguchi in
the Twenties

IT IS NOT CLEAR IF NOGUCHI FULLY AGREED WITH THE CONCEPTION OF THE viruses that Flexner had tried out in Japan and announced in his address before the AAAS. While Noguchi continued to defend his yellow fever organism he quietly pursued his own ideas on the subject of viruses. Noguchi's position at the institute, with Flexner's firm support, formed a base from which he could try to divest both his personal and professional life of past burdens and incongruities and work his way out of his personal and professional isolation. Whether he agreed with Flexner's concept of viruses or not, Noguchi felt safe in the assumption that his *Leptospira icteroides* was accepted by reliable scientists.

It was inevitable that the security that relied on ideas which were as mistaken as those held by Flexner would have a limited life span. The remaining years of Noguchi's life may be seen, therefore, as a self-contained drama: Chiwaki's original challenge when Noguchi first left Japan—that he demonstrate whether he was a man or a moral imbecile—was not yet resolved, but it could be reworded. Could Noguchi use his seemingly stable position to achieve maturity as a scientist and a man well enough to correct or compensate for failures along the way? For, in addition to the opportunity, there were other dimensions. Noguchi's ties to his Japanese past, both hostile and benign, were fading; and Time, with all its healing benevolence, was more clearly finite. What, then, could Noguchi put out of his life during these years of the twenties, and what constructive in-

fluences did he try to revive from the past or seek out from his present surroundings?

The interrelationship between his professional and personal life, which
Noguchi had tried so hard to keep separate but with only superficial success, still limited his capacity to change. A "delicious" Noguchi story which
was well known around the institute, illustrates the communication gaps
that existed, in this case between Noguchi and Murphy. It was a common
practice for staff members to pack their winter clothes and store them in
the institute basement during their summer holidays. Both Murphy and
Noguchi had Louis Vuitton trunks, very expensive French luggage with the
distinctive *LV* initials all over the exterior covering. Dr. Murphy was a
handsome man, tall and athletic in appearance and bearing, and proud of
his looks. His suits were made of elegant fabrics, tailored in London — tweeds, flannels, and full-dress formal evening wear in the height of
British fashion. Neither man seems to have known that the other had an identical trunk. During one summer, Noguchi came in to get something from his
trunk. When the first Vuitton trunk he saw didn't open to his key Noguchi
broke the lock, and pulled out all the unfamiliar contents. One story has it
that Noguchi distributed Murphy's clothes among the janitors, but Dr.
Murphy's son says that Noguchi threw them all out. When Murphy returned
to find his trunk empty, he was appalled. Noguchi did not appear to
Murphy to be too upset over the episode and made no offer of restitution.
To Murphy, the idea that anyone could have so little recognition of his
elegant taste and style that he would make no effort to learn who was the
owner of this finery was incomprehensible.[1]

Embarrassment had often been a motive for Noguchi's behavior,
whether it came from a current faux pas or from revelations of his crude
beginnings. When he learned of his aortic insufficiency and enlarged
heart, and recognized its cause, his anxiety lest it become more widely
known was clearly manifest in his concealment of the condition.

When he had to face the fact of his ultimate rejection in Japan, Noguchi
knew that the reasons for his rejection had nothing to do with basic morality or with his merits as a scientist. The prejudices that had been established
on the basis of his origins may have been rationalized in Japan by his
behavior during his Tokyo years. The fact that Noguchi was not a member
of the top echelon group from the Rockefeller Foundation, and was not
even in touch with this group in Japan, conveyed the message to the status-
conscious Japanese that he held less than top rank in the United States.
How much less, they did not know; but all of Flexner's name dropping and
laudatory remarks fooled no Japanese, accustomed as they were to mean-
ingless observances of this kind. Noguchi was always aware, too aware, of
what people thought, how he was dressed, how he was perceived, and what

status he was granted. And although he had been made a fool of by going through the motions of promising help to the Japanese with their publications without knowing that Flexner had already accepted their papers, Noguchi had made his accommodation to such embarrassment and tried to rise above it.

Noguchi's state of depression was known to all his friends in the Japanese community of New York, but it was apparently not noticed at the institute. Whether Flexner was aware of the magnitude of the pressures that had been put upon Noguchi as a result of his return trip to Japan, and whether this played any part in Noguchi's appointment to the yellow fever commission, is not clear. It was the opportunity that Noguchi had been seeking, and although he stepped into it in the highly vulnerable condition in which the events of the previous several years had left him, he had responded to it with the discovery of an organism that met the requirements of Flexner's theory of viruses. Noguchi was now irrevocably tied to the institute and began to turn away from Japan.

When Noguchi returned from Ecuador at the end of November 1918, his wife met him with the news of his mother's death. A cabled message had come to the institute and word of its contents had spread; but when he came in to start work on Monday, the morning after his arrival in New York, he showed no reaction to the news. Noguchi's colleagues could not join the pretense, but when one offered him condolences he was met with the rejoinder, "She is not gone."[2] Noguchi delayed making any response to Kobayashi for three months and then wrote the flowery speculations on the meaning of life and death that he knew were expected but added: "Having been prepared to face such an unhappy event, I consoled myself by saying, 'There is no help for it.' I remember very clearly my mother saying, when I parted from her three years ago, that she knew she would die before long." He thanked Kobayashi for caring for his parents and said: "I am sorry to hear of my father's illness, but since there is no special remedy, please give him whatever he wants."[3] His father's drinking had produced some cirrhosis of his liver.

Noguchi's responses were formalities and were in no way an expression of his own feelings. The fidelity with which he had met his obligations was the criterion by which he had been taught to judge himself, and by which he expected others to judge him. He knew that he had done all he could to fulfill his mother's ambitions, and that there was nothing he could do for his father. Any deeper feelings about what these people had meant to him, both the loving and hating that may have been stirred up by these new realities, were privately perceived and controlled before Noguchi would say a word about the events themselves.

Dr. Miyabara, who had lived with Noguchi in New York in 1906-07, and had witnessed his weeping in Japan in 1915, had promised that he would respond to any call from Inawashiro. He was as good as his word; he attended Oshika as she died, and he wrote Noguchi at length of her death and funeral.[4] Eckstein writes that Noguchi carried this letter in his pocket until it was frayed, but he never answered it.[5] Noguchi trusted Miyabara, who was not one of the despised elite, and knew that he did not have to respond to a real friend with platitudes. But the passing of his mother cut a major tie to Japan. He confined his contacts with Japan to his *onjin*, the people to whom he really felt an obligation. In this respect he remained deeply Japanese to the day of his death.

Although the summer following his triumphant return from Peru in 1920 was not at all the extended rest that Flexner had ordered, Noguchi suddenly found time to write a long letter to Kobayashi on August 24, followed by one equally lengthy the next day, mailing them both in the same envelope.[6] Kobayashi had retired from teaching and was busy with two projects; he was preparing a biography of Oshika (which was a thinly disguised recital of Noguchi's achievements and of Kobayashi's own role in making them possible) and he was organizing the school he had designed, *Nisshinkan*, for the proper education of Japanese youth. Noguchi addressed himself to both these projects, going over letters he had received from Kobayashi during the nearly two years since Oshika's death. Noguchi had not written Kobayashi since his letter from Peru in April, and that was only the second letter since his mother's death. Once he got started on the subject of his mother, however, Noguchi seems to have been unable to say all that he wanted to say or to pull himself away from the subject. He mulled over the hardship of her life and the importance of a mother's love to the development of a child. Speaking of Kobayashi's description of the services on the anniversary of her death, and of the tablets in her memory, Noguchi asked, "For how many generations have my ancestors been given death names? I imagine that the family records of poor people who cannot perform memorial services are lost from the temple." He spoke of his success having been, in large measure, the result of Kobayashi's help (he seemed to be supplying Kobayashi with quotes) and he talked of his plan to edit and publish the letters that Kobayashi had written him.[7]

Ruminating over his relationship with his mother and with Kobayashi, as though time and success had combined to give him the courage to face his mother's death, Noguchi fabricated a version of the past on which he could build a future. Born of a family whose records in the temple were tended with respect by succeeding generations, the beloved child of a mother who had worked so hard for him, and growing from that maternal warmth, his character had been cultivated by the benevolence and wisdom of his teacher, Kobayashi. And now that these loving parental influences had brought him to his current position, Noguchi was prepared to repay his obligations by helping to make opportunities available to those less for-

tunate than he had been. It had taken two years before he could venture
even this dubious accommodation to the fact of his mother's death. He can
now be seen to mourn for the parents he never had rather than face the bitterness on which he had climbed to the pinnacle on which he now stood.

Although his ramblings so often sounded synthetic and devised to please
his old teacher, Noguchi yearned to be both a man of character and the
beloved child he had never been. As he emphasized his debt to Kobayashi,
he tried to repay it in a manner that suggests an attempt at transition from
the irresponsible, bragging, and begging child to an adult who met his
obligations. In spite of the disorder of so many facets of his life, Noguchi
had kept all of Kobayashi's letters. His objective in wishing to publish these
letters was to show what spiritual guidance had accomplished for him, and
to recognize the bonds of love that existed between his teacher and himself.
Noguchi wanted to believe his own sentimental words. When he received
the John Scott medal, with an award of $800, in January 1921, he sent
1,000 yen—more than half of it—to Kobayashi for *Nisshinkan*.[8]

Nor was this only a transitory mood. Noguchi continued this kind of
support from this day forward. His wife was aware of his wishes and
demonstrated her knowledge of their meaning for him in the substantial
contribution she made from his meager estate after his death.[9] And he
continued to discuss the deatils of *Nisshinkan* in his letters to Kobayashi
until he departed for his final trip to Africa.

One of the first things Noguchi did on returning to New York from
Shandaken, at the end of the summer of 1920, was to ask that Miss Tilden's
services be assigned to him on a full-time basis, and Flexner granted this
request.[10] Starting with this change, when Miss Tilden took charge of the
administration of Noguchi's laboratory, things became a good deal more
orderly. In April 1921. he transmitted a paper that reported the work he
had done at Woods Hole in 1916 when he searched for spiral organisms in
shellfish.[11] He had set the work aside at the time, having been too depressed
or too preoccupied with the Weil's disease problems to write a report.
One of the first things Miss Tilden had done was to clean up the
laboratory, identifying the hundreds of specimens and getting them labeled
and stored in an orderly manner.* One can imagine her coming upon
the half-finished work from 1916 and nudging Noguchi to get it done and
written up. Now the search for spirochetes became a recreational activity
for everyone in the laboratory. In another paper on the subject that
Noguchi presented in April 1922, he referred to material having been
found by Miss Tilden at Woods Hole; by himself in the Catskills; by Dr.
Cross in the stomach of the sheep, ox, and cat; and by Mr. Klosterman and

---

*Miss Tilden says she cleaned up the laboratory on various occasions, and Dr. Flexner once
suggested that Noguchi be kept out of one tidy room so there would be one place to show
visitors.

Mr. Farnan in samples of fresh and salt water from the vicinity of New York City.[12] It appears that everyone was bringing in spirochetes from areas which, except for the mammalian stomachs, suggest pleasant weekend or vacation spots.

Noguchi was still planning to extend his efforts to transmit yellow fever through mosquitoes, but he was still confident of his previous work. Stokes had gone to West Africa with a commission to plan future yellow fever investigations there. When Stokes returned to England and wrote that he had not been successful in identifying the serum of convalescents in Africa with those Noguchi had found in South America, Noguchi did not regard these negative reports as significant.[13] Early in 1921, the International Health Board of the Rockefeller Foundation sent its first trainee to Noguchi's laboratory to prepare for this future work, indicating full faith in his yellow fever organism and in his methods.

The first trainee was Howard Cross, an MD from Johns Hopkins, who was soon followed by others. In spite of their advanced training in some of the country's best medical colleges, Noguchi was able to form friendships with these men, and the atmosphere of the laboratory was relatively calm, open, and friendly, perhaps in part because it was based on adoration of Noguchi. But this spirit could not last. When the foundation sent Dr. Cross to Vera Cruz during an epidemic of yellow fever there, and Cross died of yellow fever, the entire institute was upset. Again Noguchi withheld his response. On the surface he appeared to accept the explanation that Cross had not taken his preventive shots according to the prescribed routine, but Noguchi could not avoid a feeling of responsibility.

Noguchi had really liked Cross and had even grown fond of Mrs. Cross, who often came to the laboratory with her husband when he returned in the evening. Both Noguchi and Miss Tilden kept in close touch with Mrs. Cross during the first year of her widowhood, and Mrs. Cross feels that their support enabled her to survive and start her new life alone. She entered nurse's training, yet seems never to have heard that Noguchi's yellow fever work was discredited even though she rose to high administrative rank in the nursing corps of the U.S. Army and served throughout World War II. Mrs. Cross has adored Noguchi ever since. She remembers that people at the foundation almost smothered her with kindness, until Miss Tilden stepped in and told them that they were stifling her, saying "Ollie [Mrs. Cross] has a lot more to her than she is being given credit for." Thus, the foundation staff members began to encourage her wish to follow her husband's medical career by entering the Johns Hopkins school of nursing, whereas before they had been insisting that she wasn't strong enough. Both Noguchi and Miss Tilden stood by Mrs. Cross throughout her training; Noguchi came to see her on a number of occasions.[14] He wrote her two letters, which she saved all these years, and gave her thoughtful gifts, but it was a year before he spoke to her directly about his own feelings or his recognition of hers.

Noguchi's first letter to Mrs. Cross was written just after the first anniversary of her husband's death. He tried to recommend to her the ways of dealing with sorrow that he had been taught:

> The irony of it is that no matter how much one wishes, he is not able to share that suffering of the one for whom he most sympathizes! The secret of human life on this earth is perhaps to look around and realize that there are greater sorrows and miseries than that one is actually confronting. What a beautiful world this turns out to be to look out of windows and see only the purest white snow covering over every earthly thing — there lies the greatness of nature and the power of fate — and the evidence of infinity.[15]

Noguchi wrote her a second letter from shipboard when he had decided to make one more effort to establish the correctness of his yellow fever findings by going to Brazil, where the Cuban opposition had been given more attention than elsewhere. Assuring her that he was still concerned, Noguchi implied that he was taking this trip to wipe out any possibility that his error could have been resonsible for Dr. Cross's death. "I don't forget easily and I remember all things and events that moved or touched my inner self or soul. . . . You see, I have taken the field myself as I don't think I can find a more specialized person for this particular work."[16] Several times in the letter Noguchi urged Mrs. Cross to write to him. He had been deeply moved by this loving little bride of his friend, so different from the kind of woman on whom he inevitably found himself dependent. Tiny, sweet, courageous, and yet unrealistic enough to spend her lifetime in denial of the facts surrounding her husband's death, she had much that was attractive to Noguchi. Certainly he found it possible to open his heart to her just as he was trying to open himself to other personal relationships and new ideas.

On the professional side, Noguchi gave his first expression to new lines of thought in a lecture on yellow fever at Northwestern University at the invitation of Dr. Kendall. He was trying to reclassify various disease organisms that had all been grouped into the virus category for want of better understanding. Dr. Austin Kerr, who later became intimately involved with yellow fever work, was a student at Northwestern University School of Medicine at that time. He wrote: "The subject of the lecture was a very erudite discussion of *Leptospira icteroides* and its place in the world of living things. As I recall, Noguchi held that the organism was neither plant nor animal, but belonged to a special group that he proposed to create. What my professors thought of this concept I never heard, but I, for one, could not have cared less, especially when delivered by a little Jap with big glasses who spoke with such an accent that he was very difficult to understand. (At age 20 I was very much a WASP)"[17]

Noguchi had been drawn to this kind of thinking as he contemplated a revision of his treatise on spirochetes. In that work, which he had presented in his Harvey Lecture in 1916, Noguchi had discussed evidence and opinions of such early workers as Schaudinn that spirochetes were a form of protozoa.[18] Noguchi never set forth any hypothesis on protozoa, but the many times he tried to learn about them and draw them into his considerations suggest that he connected them in some way with this concept of viruses being neither plant nor animal, but a special category in between.

Noguchi arranged for Miss Tilden to spend the summers of 1921 and 1922 at Woods Hole to take the courses in protozoology for which she was so inadequately prepared; but she wrote of this time with the fondest of memories:

> We exchanged frequent letters during these two summers, not only about spirochetes, but about the protozoa, for he wanted to be kept informed about the developments in protozoology, too, and his comments were both illuminating and entertaining. You can imagine what it was to study the protozoa and the spirochaetes under such conditions; few students are fortunate enough to acquire information under the influence of that kind of intellectual stimulus.[19]

Miss Tilden's influence on Noguchi was equally apparent in the new *Laboratory Diagnosis of Syphilis*, which was published in 1923.[20] Although only Noguchi's name appeared on the book, it was a conspicuously more orderly presentation than his previous editions had been and the use of English was markedly improved. Miss Tilden says that she and Noguchi intended to extend this work in serum diagnosis of syphilis after her graduate work in chemistry was completed, a plan that might have justified her receiving no credit in print for work on this manual. She asserts that Noguchi was unusually generous in the matter of his acknowledgment of her contributions to his work.

Noguchi's relationship to visitors from Japan also reflected his efforts to adapt the forms of behavior from his cultural heritage to the Western culture, of which he was struggling to become a part. A young man, Hata (no relation to the Hata of Kitasato Institute), was warmly received by Noguchi and was entertained frequently in his home. When Hata finished his work in 1919, Noguchi invited him to celebrate at dinner at his favorite Japanese restaurant. Mrs. Noguchi joined them, taking an active part in the conversation as well as the drinking. Here Noguchi acted the role of the "big shot." Issuing orders in a loud and lordly fashion, he told the waiter to bring one of every dish on the menu, and bottle after bottle of *sake*. When Chiwaki paid a visit, however, things were altogether different. Chiwaki came in the spring of 1922, and Noguchi used his visit as an opportunity to repay his obligations to a friend and benefactor of his early years. Noguchi now played the role of the efficient, genteel host.[21]

Chiwaki was traveling partly in the interests of Japanese dentistry, and

Noguchi undertook to arrange professional contacts for Chiwaki in the United States as well as for sightseeing and entertainment. He gave a dinner at the Nippon Club to which he invited Flexner and other distinguished New Yorkers, writing Flexner in his invitation: "I have never attempted to give a dinner before, and I am anxious that this first experience should be successful."[22] He arranged "through Secretary Hughes" for Chiwaki to meet President Harding, to confer with the dentist who happened to care for both Harding and Rockefeller, and to visit the dental schools at the University of Pennsylvania and the University of Chicago under the most flattering auspices.

Chiwaki's own version of this visit adds the personal flavor of Noguchi's efforts. From New York Chiwaki wrote: ". . . I was taken immediately to the first class Hotel Plaza, on 59'th Street. . . . Dr. Noguchi knows everything about this country and was waiting with a program all set up for me, declining a lecture for the British Society for Public Health. . . ." And from Chicago: ". . . When I think back, during the thirty-eight days since I arrived in the United States, there was not one day when I did not see him. . . . When I left New York, Dr. Noguchi gave me a Panama hat which had been given to him by the President of Ecuador when he discovered the organism of yellow fever. He also gave my wife a gold watch." Back in Japan Chiwaki told of a farewell conversation when he left Noguchi in Chicago: Chiwaki had told Noguchi that although he had helped Noguchi when he was young, on this visit Noguchi had done so much for him that Chiwaki wished to cancel Noguchi's obligation to him. Noguchi was unable to speak at first; he just stared at Chiwaki, then burst into tears, saying: "What is this all about! Although I have been here twenty-two years and may not be very bright, I have not become such a Yankee that I would forget moral obligations. I still have my Japanese spirit. You say that our obligations are canceled, but an obligation is nothing to make deals with and I wouldn't think of such a thing. Please forget it yourself, and continue to call me 'Seisaku.' "[23]

Eckstein's version of this trip adds that after all the palaver and fine arrangements, which were impressive beyond question, Chiwaki reported that he had to pick up the tab for expenses.[24] If true, Noguchi was to this extent still the Seisaku whom Chiwaki had known in earlier days. Even though Noguchi had justified Chiwaki's confidence in him, and even though he was justifiably proud of the style in which he had arranged Chiwaki's schedule in visiting the United States, Noguchi was not ready to be cut loose from his obligation either materially or in spirit. His isolation was still not breached.

After bidding farewell to Chiwaki, Noguchi went up to Shandaken where, for the first time in several years, he was able to stay without running back to New York to check up on things. In his letters to Flexner he spoke of "old times" and of having watched Flexner's sons grow up from babyhood. Noguchi wrote in lavish praise of Miss Tilden's progress at Woods Hole, and of his own mood he said: ". . . everything up here is lovely. . . . I found myself a better fisherman this summer having caught quite a number of speckled trouts in the brook behind our house. Usually I catch them at night, paint them in the morning and then eat them for breakfast—a regular succession of events. My brush and canvasses do not give me any better satisfaction except I spend many hours at times. I will show you a few grotesque products in the fall."[25]

In the special summer of 1922, Noguchi was trying to write a final monograph on yellow fever, but he was not ready to believe that it was safe to abandon the vaccination and mosquito eradication programs in certain areas. He wrote the foundation that "an expert should make a personal visit . . . to examine the official records and gather whatever unofficial facts bearing on the disease are obtainable."[26] Although Noguchi had not formulated a clear decision as to the future course of his work, he appeared to be contemplating a major change.

Okumura speaks of Noguchi's spending one summer during the early twenties at Shandaken "studying high school mathematics and other such subjects," which he had missed in his haphazard education. Quoting from some unidentified source and time, Okumura reported that Noguchi wrote to a similarly unidentified person in Japan: "I have been increasingly aware that I am not sufficiently prepared for research. When I read journals of physiology and others as I have time, I am surprised at the rapid progress in many fields. Because I was locked up in the vague and inaccurate field of microbiology (my) progress was stopped and I was far away from some fine work."[27]

Noguchi had never spoken of such deficiencies in the past. If this recognition may now be taken to express his interest in going beyond laboratory techniques and the identification of pathogenic organisms to gain some understanding of basic biological processes—how a microorganism is created, nourished, modified, and reproduced; how organisms are related to each other—it suggests that his old interest in classification went deeper than the establishment of a hierarchy of living things only for reasons of convenience. Rather late in his career, Noguchi was approaching the knowledge that basic understanding of biological processes required some comprehension of their physical and chemical characteristics and behavior. In spite of his having made his flashy reputation through the discovery and manipulation of specific disease-producing organisms, if he was to achieve something of profound importance as a scientist, it was the road to understanding of the basic sciences that he must follow.

Noguchi could hardly write his unnamed friend in Japan (whom we guess

was Chiwaki) that he was broadening his knowledge of his own field
through a person who was not only a mere technician but also a woman.
But in Miss Tilden Noguchi had found a collaborator, a teacher, and an
adoring student. Although Noguchi was not yet ready to acknowledge Miss
Tilden as a co-worker, he was accepting a growing dependence on her and
was able to recognize her role in his work. She started graduate work at
Columbia University right after the summer of 1922, working in chemistry
because Noguchi believed it to be the key to the future of microbiology; if
she were to become a trained chemist, they would do great things
together.[28]

Returning to his work in New York in the fall, Noguchi took the unprecedented (for him) step of asking Carrel's assistance in the application
of new methods to his investigations. By the spring of 1923 we find Noguchi
reaching out in still another direction. He asked Mrs. Flexner's help in getting her husband to sit for a young Ecuadorian sculptor, saying: "I have
seen a good deal of his work (it has recently been exhibited, by the way)
which seems to me remarkably good—in particular some portrait busts of
prominent Ecuadorians whom I met in Ecuador."[29] Dr. Flexner wrote of
this effort in his diary:

> N appealed to me and brought the young man, a small dark in-
> dividual, to see me in my office. I explained that I did not have the time
> to sit, and was met by the proposal that he work while I was at work. I
> would give him an hour or two in the morning before interviewing
> anyone, and he agreed to finish in a few days.
> The necessary materials were brought in and the young man began.
> He would run quickly back and forth between my desk and put on or
> take off or remould the clay. This went on for days and days. It was a
> hopeless undertaking. Finally in desperation I appealed to N. One
> afternoon late he and Schmidt [Inst. photographer] came into my office
> accompanied by [illeg.] who carried a pail of clay, and between them
> they tried to get some semblance in the strange bust that the Ecuadorian
> moulded. They worked for an hour, I should say, and then replaced the
> moist wrappings. I was all agog when the sculptor unwrapped the figure
> the next day. What would happen? Nothing happened he seemed not to
> notice any change. The sittings were discontinued.[30]

Flexner intended to write a history of the Rockefeller Institute, and it is
to this circumstance that we are indebted for such reminiscences. There is
no word in Flexner's notes, from which this report was taken, that he
recognized Noguchi's anxiety in the earlier years, or that he was aware of
any desire and effort on Noguchi's part to make a new beginning. The portrait bust seems to have been one further effort by Noguchi to express his
gratitude to a man for whom he felt affection and respect. It can never be
known, of course, what there was in the creation of this bust by the young
Ecuadorian that drew such contempt from Flexner.

Hideyo Noguchi and Simon Flexner. Hideyo Noguchi Memorial Association

Noguchi's remarks at the 1914 dinner for Flexner may have been designed to convey what he knew Flexner wanted to hear, but the meaning of
these remarks to the two men was not the same. Their importance to Flexner may be judged by his dealings with de Kruif; to Noguchi they seemed
to mean the kind of atmosphere that was growing in his own laboratories
during the early 1920s.

A few months after the episode with Flexner and the sculptor, Noguchi
received news from Japan that further cut his ties to his home, and which
must have affected him more profoundly than he ever admitted to anyone.
Noguchi kept silent about the news for three months, and it is only through
the most indirect evidence that we can recognize that the news disturbed
him at all. On June 24 his father had a stroke, and on July 3 he died.[31] We
first learn of his father's death when Noguchi wrote Kobayashi in response
to the news of the great earthquake in Tokyo in September of that year. "I
was planning to write you after you informed me of the death of Jōfu but I
was tied up academically and in general and could not find a chance to
write until today's bad news."[32] This is not only Noguchi's sole reference to
his father's death, it is also the one time he used the "castle-father" appella-
tion—Jōfu—and we find no occasion on which Kobayashi (who had coined
the term for the family) ever used it for the despised Sayosuke. If Noguchi
did not entirely share the negative feelings that his mother and Kobayashi
had for his father, he had never felt free to say so. It now appears that
Sayosuke's death affected Noguchi in a more complicated way than his
cold words implied, but again the evidence is meager and slow to appear.

Following Noguchi's suggestion that an expert be sent to South America
to evaluate the yellow fever situation there, a trip was planned for the fall
of 1923, and Noguchi wrote his second letter to Mrs. Cross from shipboard. He began this letter with an apology: "It was due to a certain indisposition through which I was completely disabled to write for a long
time."[33] As he followed with all his assurances to Mrs. Cross that he was a
man who met his obligations, the words sound like those he had written to
Flexner following the death of his laboratory assistant, Steve Molinscek. Is
there any implication of atonement to these men in Noguchi's pursuing as
far as he could his investigations of the diseases that had caused their
deaths, Cross of yellow fever and Molinscek of Rocky Mountain spotted
fever on which Steve had been working during Noguchi's illness? The
record is not clear. But as Noguchi spoke of these things following his telling of the "indisposition" that had kept him from writing for so long, he
leaves no doubt that the indisposition was caused by an upsurge of feelings
on hearing of his father's illness and death. His father was the primary man
in Noguchi's life whom he had failed in meeting his filial obligations, no
matter what difficulties and resentments had prevented him from doing
so.

Although we are not to find even as indirect a reference as this one again, Noguchi surely referred to this loss unconsciously some months later. In July 1924, he attended a conference on tropical diseases and wrote Flexner from the Myrtle Bank Hotel in Kingston, Jamaica, dating his letter *June 22, 1924.** His letter began, "I arrived here yesterday morning with the second batch of men who sailed from New York on *July 16th* . . ." and continued with a list of the many "notables" with whom he had traveled.[34] It is obvious that this letter was written on July 22 rather than on June 22. Stretching the point to search for some reason for his making this error, it may be found in the fact that the period of this voyage, with all the "notables," was the anniversary period of his father's illness and death.

Regardless of how such tenuous interpretations might hold up if the full meaning of Noguchi's words were known, it seems reasonable to consider that his associations with these notables on the anniversary of the death of his pathetic father (perhaps loved with a mixture of shame) did not pass unnoticed in his mind. His tones were softer for a while. In November 1924, he wrote a brief note to Flexner: "It was very kind of you to send me nice words about the news in *Science*. Of course you are responsible for that all and I want to thank you for giving me the opportunities to receive them. But, sometimes they are cause for worriment lest I may not make good at the end."[35]

With the death of his parents, Noguchi was freed of the worst of the ambivalent obligations of the past. His debts to the friends who had shown him trust, respect, and affection now became sustaining rather than shameful reminders of dereliction. His relationships with associates in his work had become more open and even trusting, and his life with his wife had become more peaceful. Although we know little about these two people as a married couple, we have seen indications that there was much more to their life together than the rounds of drunken battles on which the institute gossip flowered.

No less significant was the evolution of Noguchi's relationship with Miss Tilden. It may have been a coincidence that the demands of work in his laboratories increased so greatly that he asked for her full-time services immediately after he began his real accommodation to the death of his mother. But coincidence or not, Miss Tilden was the strong, supportive woman who could make a satisfying career of her own as she helped him, in contrast to the woman who had been his mother but who had stimulated pity, guilt, and even rage commingled with the admiration and gratitude her help had demanded of him.

---

*Italics mine.

20

New Work and New Critics

Nature never deceives us, it is always we who deceive ourselves.
—Jean Jacques Rousseau, *Émile*

BETWEEN THE SPRING OF 1924 AND THE EARLY MONTHS OF 1927, NOGUCHI did some of the best work of his life even though it was a period during which a growing mistrust of his yellow fever work was continually battering the optimism of his new beginning. He made one more expedition on behalf of *L. icteroides*, but directed his energy as much as he could to a wide range of problems that were related to his own peculiar interests. Noguchi worked with *Leishmania*, which are protozoa causing several forms of a tropical disease generally known as leishmaniasis; he studied certain flagellated protozoa that infect plants and insects; and he made fundamental deductions regarding the structure and motility of spirochetes. He also resolved basic questions regarding the relationship of two forms of a disease known only in Peru, Oroya fever and verruga peruana, establishing in the process that a certain insect was the carrier of the organism causing the various forms of the disease. He still worked, without success, on Rocky Mountain spotted fever and undertook studies of trachoma that he left incomplete and that created a new source of controversy.

Throughout this very fertile period of Noguchi's work, the conviction of the Cubans that his yellow fever work was erroneous could not be dispelled. Not only did the mounting body of conflicting findings distract Noguchi from all his other interests, it also affected the reception of his other work.

232

His detractors were more cautious in their acceptance of anything he published, and his supporters became increasingly active on his behalf. Distinguished scientists began to preface their comments on Noguchi's work in other fields with disclaimers of opinion concerning his yellow fever work. Flexner, on the other hand, wrote: "Considerable hostility has always been shown by the Cuban Department of Health to the *Leptospira icteriodes* view of the origin of yellow fever. The circumstances are unfortunate and personal, but unavoidable. However, this feeling has led to numerous unsuccessful efforts to discredit Dr. Noguchi's work."[2]

Although Miss Tilden remembers that Noguchi regarded the yellow fever work as having been completed at this time, Noguchi was uneasy about the constant reappearance of the disease in Brazil. Public health workers in that country, strongly influenced by the Cubans, had undertaken a mosquito eradication program that had greatly reduced the incidence of yellow fever in Brazil but had not eliminated it. The Brazilian health workers had never encouraged the foundation's offer of assistance, but they finally did accept it in 1923. In November of that year, Noguchi and one of his trainees, Henry R. Muller, embarked for Brazil with a cargo of microscopes, sterile glassware, incubators, iceboxes, guinea pigs, rabbits, four Maccacus rhesus monkeys, three baboons, fifty white rats, and fifty mice, the animals all in "mosquito-proof crates."[3]

They were received with courtesy and elaborate facilities were put at their disposal, and once again Noguchi bloomed in a Latin environment. He sought out a teacher of Portuguese and soon was not only speaking the language but was writing Portuguese poetry as well.[4] And, once again, Noguchi thought that he had succeeded in identifying his *Leptospira* and had convinced the local workers that it was associated with yellow fever: "The work was so planned that I had the Brazilian bacteriologists to do the tests themselves." He wrote both Flexner and Russell at the foundation that he had conducted a demonstration before the entire faculty of the medical school, and the results were clear-cut.

> Thus it was regarded by all who participated in or witnessed the reaction that *L. icteroides* is specific for yellow fever . . . Now they are all absolutely convinced of the presence of *L. icteroides* in the Brazilian yellow fever cases. They also realize why they failed in their own attempts previously. They were untrained experimenters and it would have been a wonder if they had succeeded.[5]

Just as Noguchi had annoyed the Cubans by pointing out the faults in their technique, for which he thought they were delighted and grateful, he may well have created the same irritation in the Brazilians and similarly misread their reactions. "I came to Brazil partly because I wanted to show them the possible sources of failure and how to get a positive result. I believe I have succeeded in both."[6]

An important element in Noguchi's ability to convince others, to whatever extent they were convinced, did come from his relatively sophisticated experience with the dark field microscope. The next generation of physicians were trained to use this technique as part of their routine medical education and they did not regard it with the awe in which Noguchi and others of his generation held it. Beyond that, his persistence and unquestioned facility in finding whatever spirochetes were indeed present were always an impressive spectacle. But again, his own conviction and need to believe made it possible for him to carry others along with him to accept his interpretations, even when they may have had second thoughts after his departure. In Brazil, moreover, Weil's disease was no better known as a clinical entity than it was in any other American country.

The apparent success in Brazil was the last straw for the Cubans. At a conference on tropical diseases in Jamaica during the following summer (1924) the senior Cuban public health officer, Dr. Aristedes Agramonte, the sole surviving member of the original Walter Reed Commission of 1900 in Cuba, led an attack on Noguchi's yellow fever organism and the whole program of vaccination as it was being offered by the Rockefeller Foundation. In a prepared paper, Agramonte reviewed the original work done in Cuba, emphasizing the mosquito eradication program and maintaining that such a program was still the only proven method of controlling the disease. "The protective vaccine at present used in yellow fever is one obtained from inoculations made upon animals with Professor Noguchi's leptospira. I must declare that this *Leptospira icteroides* presented as the causative germ of yellow fever . . . fails to fit in with etiologic facts relating to the disease in man, as we know them to be." Sarcastically referring to Noguchi as "the savant," Agramonte pointed out the numerous failures of others to confirm Noguchi's results and the fact that these failures had been explained by Noguchi as the result of faulty technique. To this last criticism, Agramonte took special exception. He described what he considered were the significant clinical differences between yellow fever and Weil's disease, and he suggested that it would be far less dangerous to use human volunteers for truly objective experiments, as had been done in Cuba, than to expose entire populations to the false security of a useless vaccine.[7]

Agramonte turned out to be correct in much of what he said. However, his suggestion to use human subjects for experimentation on yellow fever and his apparent intention to be insulting to Noguchi (perhaps to the foundation as well) weakened his case. Besides, others were not as sure as he that yellow fever and Weil's disease were so readily distinguishable. As the story of Agramonte's attack has been repeated for fifty years among yellow fever workers, it is commonly told that "Noguchi squirmed and took it."[8]

Almost two weeks passed before Noguchi wrote Flexner about it, and we may assume that he had been quite angry but had taken time to cool off. In his usual style, Noguchi wrote many pages about other things before he approached the subject, and then he dealt with it as casually as he could. He wrote that Agramonte's objections were very unreasonable and he was not sure whether "these Havana men are really interested in scientific discussion or not."[8]

The published transactions of this meeting include remarks that were made in discussion and which suggest that doubts had appeared among those who had previously seemed to believe firmly in *L. icteroides.* Dr. Henry Rose Carter, who had been in Cuba even before Gorgas and Reed and who had known Noguchi in Peru, summarized the problem well when he said:

> I accepted this *Leptospira icteroides* as the causative organism of yellow fever, not because of my own findings, but simply from the statements of Dr. Noguchi. If his statements are true, and I have every reason to believe they are, to me the conditions for belief in his results are satisfactory. I have vaccinated a number of people going into the Tropics — and I shouldn't think of allowing my daughter to go into the Tropics without being vaccinated — but that is only evidence of my own belief in the value of vaccination, and not evidence of such value. The instances . . . that have been mentioned are evidences *indicating* the value of Dr. Noguchi's serum and vaccine, but do not prove their value.[9]

This is the kind of support that can encourage doubts rather than erase them; it may have worried Noguchi more than Agramonte's attack. But Noguchi did not know what he could do about it and turned away to other problems. That was not so difficult for him to do since he was presenting a paper at this Jamaica meeting on a totally different problem, one that had aroused his interest during his stay in Brazil.

Several forms of the tropical disease leishmaniasis had been described clinically from various countries, but workers had reported difficulty in getting sufficient growth of the organism for laboratory study. Noguchi was intrigued, perhaps by the reported difficulties as much as by the fact that the organism was a protozoan. Working with Dr. A. Lindenberg, who was in charge of the Leishmaniasis ward in a hospital in São Paulo, Noguchi was able to cultivate the organism on his spirochete medium before he left Brazil.[10] He carried "a few strains of *Leishmania* from the Brazilian type of leishmaniasis" back to New York where he identified three species that could be distinguished serologically. His findings have since been confirmed and correspond to the classification and clinical distinctions that are accepted today.

Noguchi read a paper on this work at the Jamaica meeting.[11] Not only was it well received but, perhaps more important, it could be appraised by men who approached his work without prejudice since they knew relatively little about yellow fever and its related problems. A number of distinguished zoologists, including C.A. Kofoid of the University of California at Berkeley, Aldo Castellani of Brazil, Juan Iturbe of Venezuela, and R.W. Hegner of Johns Hopkins University, were interested in protozoa that produce disease in plants and in invertebrates as well as in vertebrates.[12] Some of these protozoa have been found in insects that feed on plants subject to infection by such organisms. It was a unique opportunity for Noguchi to learn from these men who received his offering so well and who opened up to him a field with which he had had no previous contact. It was also a welcome antidote to Agramonte's harsh words.

Noguchi's new friends shared their knowledge of *Leishmania* with him and told him of laboratory facilities at the hospital at Tela, Honduras. Noguchi already knew of studies that had been made by R. P. Strong and his associates at Harvard with material from this area.[13] Therefore, when the members of the conference were invited by the United Fruit Company to go on an excursion to some of the Central American countries on one of the company ships, Noguchi was delighted to join the other scientists who accepted this invitation.

Noguchi had long ago noted the suggestions of such early workers with spirochetes as Schaudinn that the syphilis spirochetes (as well as others) might be a protozoan, and he wanted to learn more about these organisms. He probably no longer accepted Schaudinn's suggested view, but there are hints that he hoped to find, in the study of protozoa, answers to questions that arose in his work with spirochetes. How do spirochetes achieve their motility? What characteristics might be found to relate the various insect-borne organisms to each other? Microbiologists of today, with the help of greatly enhanced technology, do not have to depend on such devious methods for gathering information about organisms. But Noguchi enjoyed the pursuit of any microorganism, regardless of the objective or the method he had to employ. He knew at this time that certain protozoa have flagella — whip-like appendages by means of which they are able to move. He also knew that spirochetes have no such appendages, but he had watched their movements by the hour and the question of their motility fascinated him.

The main body of the paper that Noguchi read at the Jamaica meeting reported his cultivation and serologic work with *Leishmania,* but he provided an addendum to the manuscript when he sent it in for publication. Here he described the flagellated organisms he had found in plants not only in Honduras but also around New York after his return, and he compared their physical characteristics with those of *Leishmania.* Noguchi described the movements of all these flagellates in such detail as could have been accumulated only through long days and nights with his microscope, reporting that there was little difficulty in distinguishing the plant flagellates from the insect flagellates. He even drew up a chart comparing their measurements.[14] Noguchi expanded this work during the next two years, published one paper under joint authorship with Miss Tilden,[15] and then suddenly diverted his complete attention to the application of his observations of these flagellates to his old interest in the motility and structure of spirochetes.

It is difficult to determine from his reports and publications just how and when Noguchi arrived at his conclusions on this subject. Presented with none of the bombast with which he had proclaimed that the pure culture of *T. pallidum* had been "proved beyond all doubt," Noguchi's hypothetical description of the motility and structure of these organisms was stated modestly but firmly. Although it created no stir in the scientific world, and for many years no one gave it credence, when the electron microscope extended the visual scope of microbiologists they found that Noguchi's description had been entirely correct.

Noguchi made some tentative suggestions along these lines in a note on "Abnormal Bacteria Flagella in Cultures" published in the *Journal of the American Medical Association.*[16] A few months later he reported his ideas with greater assurance to the Board of Scientific Directors of the institute.[17] It was only in the chapter on spirochetes that he wrote for Jordan's and Falk's textbook on bacteriology that he presented his fully developed hypothesis.[18] From this source, however, this hypothesis was noted and finally discussed in the textbook on microbiology originally written by Noguchi's old friend Zinsser, but long since taken over by others:

> *Motility:* Spirochetes are motile. They progress by sinuous and rotating movements of the body. Some of them have delicate terminal filaments. Noguchi believed that the essential structure of a *Treponema* consisted of a spring-like axial filament and a layer of contractile protoplasm enclosed in a delicate protoplast. He believed that the axial filament was a kind of intracellular flagellum. Noguchi did not illustrate the axial filament, and most observers for the next 40 years denied its existence. The first electron micrographs failed to show axial filaments but did reveal thread-like appendages resembling flagella. Subsequent studies have confirmed Noguchi's conclusions. External flagella are not found on organisms examined directly from living man or animals. However, after tryptic or peptic digestion for 10 minutes, three axial filaments can be seen in the body of *T. pallidum,* and eight to twelve in *Bor. recurrentis.* After 20 minutes digestion, the filaments are freed from the body and appear as flagella. The contractile material in the *Leptospira* is wound around the outside of the organism.[19]

How could Noguchi present, with such assurance, a precise description of the axial filament that no one could see? Under the heading "Motility and Its Mechanism" Noguchi had written: "From a study of the degeneration phenomena of several types of spirochetes the writer has concluded that the kinetic element in these organisms is the axial filament, which is a

kind of modified flagellum. The reasons for this deduction will appear in
the separate discussion of the various genera given below."[20] The difference between Noguchi's view of the axial filament, as it appears in these
discussions, and the view of those who denied its existence, came from the
interpretation put upon that which was visible under the microscope.
Beginning with the largest organism, Noguchi recorded what he saw and
made "reasoned" deductions from a kinesthetic perception of the sequence
of events taking place under his microscope.

The largest organism was the *Cristispira*, the organism that Noguchi
had begun to study at Woods Hole in 1916. He described no more than the
others had observed of its structure and movements; but his word picture
of the movement led to his first reasoned deduction:

> It is difficult to follow the shapes taken by the body in motion, but it
> seems to straighten and relax in rapid alternation. It is clear that the
> elastic wavy membrane is stretched at short intervals by the rhythmical
> contraction of the cylindrical body. In a fluid medium the only possible
> outcome of such contractions is constant rotation of the organism, the
> spiral membrane propelling the body backward or forward according to
> the direction of rotation. From the mechanical and morphological view-
> point the crista is a giant compound flagellum in which innumerable
> fibrils are united.[21]

From the *Cristispira* Noguchi proceeded to his discussion of the much
smaller spirochetes, beginning with the *Treponema*. As he extrapolated
his deduction from the *Cristispira* to *Treponema*, it becomes clear that his
interpretations came out of the accumulation of perceptions during the
years in which he had watched and, most important, made drawings to
record what he saw. Noguchi had followed these organisms through all
stages of their life cycle; he had subjected them to dozens of variations in
their chemical and physical environment and observed their reponses; he
had required that Bronfenbrenner observe them and draw pictures of all
that this impatient young man could see; and he had studied Bronfenbrenner's pictures at length. In addition to his own drawings, Noguchi had
photographed the organisms in various phases of their lives.

In other words, he was adequately prepared, by personal endowment
and experience, to approach this problem. No vacuum of chemical
knowledge stood between Noguchi and the behavior of these *Treponema*
as had existed in his interpretation of experimental manifestations of the
Pfeiffer phenomenon. No clinical judgment was involved, and no
newsworthy event could be anticipated to enhance his career or guarantee
the security of his position. And he was under no threat of exposure of
previous errors. It is significant that Noguchi presented his hypotheses only
as reasoned deductions but with such quiet reassurance. Less than two
years later he would write Flexner from Africa about a completely spurious
organism that he hoped would erase his error in yellow fever: "I have never

been so certain of being correct as I am with this piece of work on West African yellow fever."[22] Yet he bravely but calmly went on to make further extrapolations from these tiny *Treponema* to explain the structure and motility of the even more delicate *Leptospira*.[23]

Corner wrote of Noguchi's search for *T. pallidum* in the slides from the paretic brains: "Noguchi brought to the quest no new method, but only his own high-strung determination, persistence, and visual acuity."[24] The same may be said of these later motility studies with the addition of a special characteristic that was essential to his intuitive comprehension of the physical process which he described with such precision. Even before his work with paresis, Noguchi had told his young friend Akatsu that he could not delegate work to assistants as he knew Paul Ehrlich had done. He explained that Ehrlich worked "with his head" and so could evaluate the work of his assistants. Speaking of himself, Noguchi said he could not trust his head as Ehrlich could; he could trust only his hand and thus had to make his own drawings of all that he saw under his microscope. It would appear that in making his drawings, which he did with elegance and a real sense of movement, Noguchi acquired a dynamic perception of the kind of activity that would produce the movements he saw, and from such perception he could rationalize the structure required to produce such activity.

It is a grievous loss to science that Noguchi did not publish his own drawings to illustrate the process through which he made these deductions. In the Jordan and Falk textbook, he published photomicrographs of living and dead *Cristispira* at different stages, together with what he called schematic drawings "showing the structure and disposition of the membrane or crista, as brought out by dark field observations of the course of degeneration."[25] A similar elucidation of his deductions in the case of the axial filament within the body of the *Treponema* might well have illustrated his kinesthetic recognition of the simple physical phenomena. Noguchi could thus have demonstrated how he was led to the intuitive hypothesis, without which no basic scientific principle has ever been conceived.

Yet while Noguchi was so deeply immersed in this most creative of all his work, he was planning, working, and supervising the work of others in his laboratory, on Oroya fever and verruga peruana, the two phases of a disease generally known as Bartonellosis, which is found only in Peru. At the time Noguchi first heard of this disease in 1920, the opinion had been advanced that it was carried by an insect, but no one had presented scientific evidence in support of this opinion. The disease has a dramatic history that was first recorded by the Incas of Peru in their stone carvings. The manifestations of the disease are equally exotic; it is contracted only within altitudes of 2,500 to 8,000 feet above sea level and only at night, and is manifest in:

two illnesses which are strikingly different in their appearance and in their underlying pathologic processes. In addition to causing non-clinical asymptomatic infections, this organism can produce either one or both of the syndromes known as Oroya fever and verruga peruana . . . .Oroya fever, the initial stage of infection, appears as a severe, febrile hemolytic anemia. . . .The second stage of infection is verruga peruana, which appears as multiple eruptions of disfiguring, heman-gioma-like tumors. Although verruga peruana will usually follow a recognized attack of Oroya fever in 2 to 8 weeks, the infection with Orova fever may be of such abnormal mildness that the patient will have no recollection.[26]

The syndrome acquired the name Carrión's disease in 1885 in honor of a young medical student, Daniel Carrión, who had himself inoculated with material taken from a verruga patient. Carrión's objective was not to prove the identity of the source of the two illnesses, but to study early symptoms in the hope that earlier diagnosis might prevent some of its disfiguring consequences. When Carrión first developed a case of Oroya fever from which he died, he and his fellow students who had cooperated with the experiment only incidentally recognized the dual manifestations of the disease. From this time forward, Peruvian physicians generally accepted this concept of the etiology of the two forms of the disease, but the Harvard Commission that had gone to Peru in 1913 reported otherwise.[27]

In 1905, a Peruvian physician, Barton, had described an organism that he believed was specific to Oroya fever, and the commission members confirmed his observations. They found the organism (to which they gave the name *Bartonella bacilliformis*) in the blood and tissues of Oroya fever patients, but they failed to find it in cases of verruga peruana. The commission reported its conclusion that the two diseases were separate and distinct entities. In 1912, Townsend, an entomologist living in Peru, had started his own informal search for the insect vector of the organism and identified the *Phlebotomus* sand fly. Townsend's first efforts led to his observation that only persons who remained in the infectious area during the night became infected; from this observation he searched for an insect whose distribution and habits fit the geography and nighttime exposure of humans and thus labeled the *Phlebotomus* as the likely agent.[28]

This kind of background made the problem a most intriguing one for Noguchi. The association of the disease with a single man who lost his life in proving a point, the contrary opinions of the Harvard Commission regarding the dual nature of the illness but their appealing suggestion that the organism was a protozoan (which it was not), the association with an insect vector — all fit Noguchi's recent interests and his style. He secured blood samples from Oroya fever cases from Peru and was immediately successful in culturing the organism from this source. He injected it into the skin of monkeys and produced typical verruga warts at the site of the injection, and he injected it into veins of monkeys and produced Oroya fever. He then secured samples of human warts from verruga peruana cases in Lima and, similarly, produced both diseases in the experimental monkeys.[29]

Not satisfied with having demonstrated that the two diseases were, in fact, different manifestations of an infection with the same organism, Noguchi proposed to round out his study with the positive identification of the insect carrier. It was later arranged that Raymond Shannon, an entomologist, would go to Peru to collect insects for Noguchi. Miss Tilden writes of finishing this part of the work after Noguchi left for Africa: "Unfortunately these positive results were not obtained until after Noguchi's death; hence he never knew that his plan for detecting the insect carrier of Carrión's disease had proved successful."[30]

Dr. David Weinman says that because so much of Noguchi's work was being questioned at the time this work on Carrión's disease was published, many people reserved judgment on his findings. Furthermore, Noguchi's papers were written in such a confusing manner that it was difficult to follow just what he had done. Weinman was a member of a second Harvard Commission that went to Peru in the early 1930s to reinvestigate the matter, and this group confirmed Noguchi's findings. By dint of careful following out the details as Noguchi had reported them, jumping back and forth from one paper to another, Weinman satisfied himself that Noguchi had, in fact, done just what he had claimed.[31]

That it was one of the soundest pieces of work of Noguchi's career seems clear. That it did not gain more favorable attention was due, no doubt, to the obscurity of the disease and to its being overshadowed by Noguchi's problems with yellow fever. For during the whole of this highly productive period of Noguchi's professional career, the reports of other workers in yellow fever were coming in steadily, and almost all contradicted Noguchi's contention that *L. icteroides* was the causative agent.

Muller had gone to West Africa in the fall of 1925 and had found no *Leptospira* in the yellow fever cases there. Kligler had been recruited from Palestine in 1926, and after six months in Africa he, too, had found no *Leptospira*. The Rockefeller Foundation discontinued its distribution of yellow fever vaccine and immune serum, but it is not clear whether this was done because there seemed to be no further need for the vaccine and immune serum, or because of doubts that had arisen. But perhaps the worst blow of all was the report of a series of experiments performed at the Harvard School of Tropical Medicine by Theiler and Sellards. After Noguchi's return from Brazil, these men undertook a thorough investigation of the serologic relationship between *L. icteroides* and *L. icterohemorrhagiae*, the organism associated with Weil's disease, exactly as Noguchi should have done when he first returned from Ecuador.

Theiler and Sellards obtained a strain known as Palmeiras No. 3 from Noguchi. Over a two-year period they standardized all their materials, techniques, and animals, and proceeded to compare this strain of *L. icteroides* with a strain of *L. icterohemorrhagiae* that they found in Boston rats. They used the Pfeiffer phenomenon, as Noguchi had done, to determine the immunologic relationship between these two samples of organisms and reported that their reactions were identical: that guinea pigs that had been actively immunized against infection with either organism received complete protection against the other. In conclusion, they stated:

> This constitutes strong evidence of serological identity of these strains of leptospira. These results do not afford any additional evidence regarding the etiological relationship of *L. icteroides* to yellow fever, especially since the exact relationship of Weil's disease to yellow fever is by no means completely understood. One is confronted with possibilities too radical for discussion at present, the extremes being that Weil's disease and yellow fever may be etiologically identical, or that the leptospira may have no etiological relationship to yellow fever.[32]

Many years later, after many more vicissitudes, Theiler was awarded the Nobel Prize for the development of a yellow fever vaccine. In 1972, shortly before he died, Theiler was reluctant to criticize Noguchi, even though he had seen Noguchi's laboratory and had found it unbelievably disorderly and sloppy. Theiler did, however, condemn Flexner in several episodes that he related, some of which were connected with yellow fever and some with other problems. But Theiler was particularly saddened by the extreme suspicion that he heard people express in relation to anything Noguchi said or did once the questioning of his yellow fever work began to grow in its intensity. He told of Noguchi's coming to Harvard to speak on Oroya fever. When E. E. Tyzzer, who had been a member of the 1913 Harvard Commission in Peru, saw some pictures that Noguchi showed, Theiler heard him say, "This doesn't look right to me," and he heard Noguchi reply, "They are your own pictures, Dr. Tyzzer." Theiler recommended that the whole Noguchi matter be allowed to "rest in peace"; he said it was surely a psychological problem.[33]

21

Reality

Yesterday and today I am feeling the weight of accumulated years. I have not
forgotten, but time passes.

—Noguchi to Kobayashi, March 1927[1]

NOGUCHI COULD NOT AVOID FACING THE QUESTIONS ABOUT HIS YELLOW
fever work indefinitely. Both he and Flexner had expected that *Leptospira*
would be found in yellow fever cases in West Africa to support their
position, but Flexner continued to hope that other workers would
accomplish this task. Noguchi wrote Kobayashi that he had sent
investigators to Nigeria but "it may be necessary for me to go later."[2]

Meanwhile he worked harder than ever on the many projects he had
undertaken. His chapter on spirochetes was comprehensive; a well-
organized survey of current knowledge of both pathogenic and non-
pathogenic forms of the organisms. Much of this work was his own, but as
we have seen in the matter of the structure and motility of spirochetes,
Noguchi extended his work in the course of writing this review of the subject. In addition to his work on Bartonellosis, leishmaniasis, and the plant
and animal flagellates, he continued to struggle with Rocky Mountain
spotted fever and started work on trachoma.

During the years immediately following his trip to Brazil, the
fluctuations in Noguchi's moods were becoming more conspicuous. In
April 1925, while Flexner was on an extended trip through the Near East,
Noguchi wrote him in rather depressed tones:

We had, yesterday, the Directors luncheon, this being the second where we all missed you—I specially did feel perhaps more than anyone for the obvious reason. I need not remind you. . . .Now, as for myself, I have been rather slow in my work—not that I do not work hard enough but perhaps the problems are not easy ones or else I am falling in my work—Anyhow, I have no new discoveries to startle you or please you.[3]

Noguchi told Flexner that he had started some experiments in yellow fever, using a chimpanzee as an experimental animal, and had assigned the task of gathering data on the normal chimpanzee to various people in the laboratory. "Battistini has been doing blood analysis of the chimpanzee and guinea pigs for PH (sic) ranges in relation to yellow fever infection." But he was uneasy. "Somehow I cannot manage to find enough time to sit quietly and think over things calmly and reflect upon many things and phases in life. I seem to be chasing something all the time, perhaps an acquired habit or rather the lack of poise."

This casual introduction to Telemaco S. Battistini was extended in a postscript. Noguchi reminded Flexner of Battistini's background in Peru and of the two years he had already spent at the institute as an International Health Board fellow, and he asked that Battistini be appointed as his assistant for the next year. "He says he will take his family back to Lima this summer and will come back if I wanted him. . . .Of course, no assistant is absolutely essential to my work. I merely thought it will be of mutual benefit."[4]

Battistini had recently completed a study that was reported in a paper published under his sole authorship.[5] A strain of the organism that the Kyūshū group had discovered to be the cause of the so-called seven-day fever of Japan, as reported in 1918, had arrived in Noguchi's laboratory.[6] It had been identified as a member of the *Leptospira* group. Battistini had compared the serologic relationships between the "three known species of leptospira"—those of Weil's disease, of seven-day fever, and of the *Leptospira* Noguchi believed to be the cause of yellow fever—and reported consistent results with all three. That is, the antiserum prepared with each organism produced a positive reaction to all strains of that organism and negative reactions to each of the others.[7] Such results must have been most gratifying to Noguchi, but from the hindsight of today, we can see that Battistini must have made some major error: although the seven-day fever and Weil's disease organisms are now known to be related, they can be distinguished by their serologic reactions as Battistini's experiments showed. But *L. icteroides*, the yellow fever organism he believed he was using, does not exist as a distinguishable organism.

Noguchi had not told Flexner that Battistini was also interested in Oroya fever, on which he had tried to do some work in Peru, or that Battistini had promised to collect blood samples from patients with Oroya fever during the summer in Lima and bring them back when he returned in the fall. Nevertheless, Flexner arranged for Battistini's appointment, and Noguchi was admittedly more cheerful when he wrote Flexner from Shandaken in August. Now hoping that his yellow fever work would be confirmed by Muller, who had been with Noguchi in Brazil and who was about to leave for West Africa, Noguchi wrote, "I believe that yellow fever work may now be relaxed."[8]

Noguchi had taken all the equipment he required for cultivation work to Shandaken with him and had achieved some success with flagellates, which he reported with pleasure.

> When I enter the field of Kala Azar [a form of leishmaniasis] I intend to use this method to clear up some of the problems you mention in your last letter. I am really happy because I feel that I can still be trusted with devising newer methods in studying various organisms. And it is nice once in a great while to get what one goes after. I am writing this note with somewhat different mood of spirits than my last note — I am inspired and confident — full of expectations. . . .Yes. I fully realize that I am no longer a boy of yesterday, but that time flies faster with each year added to our credit or discredit of life. Perhaps a wonderful, mature and fundamental work may yet come to rescue the inevitably shrinking days, and strange thing is that I have peculiar optimism despite of many deceptive years that passed.[9]

Battistini returned with the Oroya fever blood samples in September, and together he and Noguchi succeeded in their first attempt to culture the organism. They published a preliminary report in February 1926,[10] followed by two papers under their joint authorship.[11] From that time on, however, Battistini's name no longer appears. Furthermore, he resigned from the institute as of January 31, 1926, and departed at once for Peru.[12] Later workers in Peru learned that Battistini felt that he was denied credit for the contributions to Oroya fever studies that he had made before coming to the United States, and which were published in Peru in 1925.[13] We cannot know why Noguchi chose to ignore these publications; we can only observe that Battistini's resignation was only one of several indications that an air of distrust was building up in Noguchi's laboratory.

Shortly after Battistini's return from Peru in the fall of 1925, Dr. Johannes Bauer arrived. He had come to New York on furlough from Peking Union Medical College and soon joined the yellow fever work in Noguchi's laboratory. Although Bauer was to play an important role in yellow fever work for many years, his influence on Noguchi is hard to appraise. He eventually became director of the Health Division of the foundation, and he was in charge of a Red Cross mission in Europe that opened up the concentration camp at Buchenwald after World War II. Bauer was clearly a man of considerable and varied talents. Of Swedish nationality, and educated in Germany and Russia, he became known as a very competent but rigid worker. He showed the same characteristics in his personal relationships.

Noguchi specifically noted that Bauer performed the blood counts in the experimental animals for the Oroya fever studies, and that he suggested changes in the culture media that resulted in improved growth of the organism.[14] When Noguchi had first asked Flexner to arrange for Battistini's appointment, he said that Battistini had been doing this work for the yellow fever studies, and it would seem natural for him to perform such tests for the Oroya fever studies in which he was so directly involved. For no apparent reason, however, Noguchi not only assigned this work to Bauer; he also chose this time to ignore Battistini's previously published work in Oroya fever, thus precipitating Battistini's resignation. We do not know why Noguchi suddenly lost confidence in Battistini. But we do know that in another laboratory, Bauer was charged with misguided attempts to produce experimental findings that he knew his laboratory director hoped to see.

Dr. Carl TenBroeck, who had just accepted the post of director of the laboratories at Princeton, to replace the retiring Theobald Smith, accused Bauer of having falsified data in work that he had done as TenBroeck's assistant in Peking. TenBroeck claimed that he had published the work in good faith, and he now apologized to the institute for having done so and offered to withdraw from his directorship.[15]

How this situation was resolved we do not know; the result, for the next few years at least, was that Flexner distrusted both Battistini and Bauer. Flexner later recommended that the foundation find someone other than Battistini to take charge of the collection of insects in Peru for the work that Noguchi had started with Oroya fever,[16] and he similarly resisted turning over work with yellow fever cultures to Bauer. "I know little about Bauer and hate to put the responsibility in his hands. . .the TenBroeck incident has not been explained. The two letters that Noguchi sent Miss Tilden and of which she gave me copies do not increase my confidence."[17]

Noguchi's behavior in the laboratory at this time was described by Dr. Charles Doan, a young man whom Flexner persuaded Noguchi to take into his laboratory to demonstrate some newly developed techniques for staining living cells.

> Dr. Noguchi was very reluctant to let me come into his laboratory. . . . He pledged me to secrecy, and said that I was not to say anything about anything that I saw in his laboratories, about his work. This, of course, I readily agreed to. We did study his material and found that the organisms moved rapidly throughout the red cells, and these I attempted to show him, but he still rejected this new technique, saying that he would rather stay with his own silver and other stains for spirochetes. Dr. Noguchi became more and more obsessed with the conviction that if he could go to Africa, he would be able to show what he had shown

in Central America. . . .I well remember when Dr. Noguchi made his farewell talk before leaving for Africa, saying that he would prove that he was right and that nothing would prevent the ultimate success of this expedition.[18]

But even while all this disturbing activity was going on in his laboratory, Noguchi continued to accumulate honors and awards. In 1924, the French government made him a Chevalier of the Legion of Honor, and in the fall of 1925 he was awarded an honorary degree by the University of Paris at the Sorbonne. Noguchi did not interrupt his work to be present when his degree was awarded in Paris, but he did attend a formal dinner that the institute held in his honor on January 15, 1926. Invitations to the dinner were typewritten on institute stationery, and Flexner wrote in pencil on the side of his invitation, "accept. Say nothing to Noguchi."[19] Perhaps there was some element of surprise.

Formal speeches in adulation of Noguchi were the order of the evening. The address that Professor H. Roger, dean of the Faculty of Medicine at the University of Paris, had delivered when Noguchi's degree was conferred in absentia, was read in translation. Among the many speeches of the New Yorkers, the friendly informal talk by Dr. Rufus Cole, director of the hospital, portrayed Noguchi as his colleagues knew him: indeed, Miss Tilden refers to Dr. Cole's remarks as describing "the Noguchi I knew." Dr. Cole teased Noguchi about his accumulation of awards, calling him "our prize medal winner. . . .I suggest that we should have a trophy room where the material evidence of all the honors that have come to Dr. Noguchi could be exhibited. What a display it would be and what a source of envy for all rival institutions! But shorn of his decorations and medals we would still be proud of our associate, whom we all regard as a brother and one to whom everyone is devoted. That Dr. Noguchi was not born in America is not his fault. We claim him unreservedly." Cole went on in this vein until he came to "the real secret of Dr. Noguchi's attractiveness. . . and why his work is so uniformly successful. . ." which turned out to be Noguchi's resemblance to the child who noted and mentioned that the king had no clothes on.[20]

Noguchi wrote Flexner the next day expressing his appreciation for years of kindness and interest in his career, all of which had culminated in this dinner.[21] Two weeks later, Noguchi wrote to Rockefeller on paper with the printed heading, "Office of John D. Rockefeller, Jr." We do not know if Noguchi went to the office to write the letter, or acquired the paper otherwise. Thanking Rockefeller for the honor of the dinner, Noguchi wrote:

I never dreamed that once in my life I was to be feted by the greatest men of the day and particularly to be seated at the same table with you. This is all like a dream to me. . . .I am not unaware that

the honor to me from the Sorbonne is really for the great name of
the Institute. . . .However, I am very happy to be picked out as one of
those whom the University of Paris deemed worthy of the compliment.[22]

Six months later Noguchi wrote Kobayashi telling of the honorary
degree and listing the names and ages of the seven other men who were
similarly honored at the same time. Perhaps even Noguchi did not realize
how truly honored he had been; the list included "British chemist Lord
Rutherford (age 55-56). . .and Russian physiologist Pavlov (age 75)."

At my Institute they celebrated this in December. Even though anti-
Japanese feelings are expressed, human feelings are equal everywhere
as well as friendship. The Japanese who are in the United States
should respect the etiquette of this country and try to live respectfully.
Then Americans will not treat them cruelly. The anti-Japanese feelings
of today were brewed by the Japanese themselves. Of course, without
being careful of language, customs, philosophy, actions, and personal
characteristics which are quite different from the Japanese, we cannot
have good relationships with them.[23]

As he continued to tell Kobayashi about his plans for the coming fall,
Noguchi spoke of visiting the African coast "if it becomes necessary," but
implied that such a visit would be for the purpose of studying trachoma
and Kala Azar. Behind these screens he set up for Kobayashi, as in the
formal expressions to others, Noguchi knew that his day of reckoning with
yellow fever was not far off. The reports from Africa were far from
reassuring. Noguchi may have been influenced by his desire to get away
from this bad news that he could neither refute or confirm, but other
personal considerations may also have played a role as he allowed himself
to be drawn into studies of trachoma. He had no doubt seen much
trachoma in Japan, and had tried to work with this disease more than once
before this time. When Eckstein went to Japan to gather material for his
biography of Noguchi, he found Inu, Noguchi's sister, almost blind from
trachoma. How far advanced her condition was when Noguchi saw her in
1915 we do not know.

Noguchi made a trip to Albuquerque in May 1926 to gather material for
his work on trachoma. He took monkeys with him and injected their
eyelids with material taken from patients there, and produced lesions that
could be passed from one animal to another. It was not clear, however,
that the disease being transmitted was the same as human trachoma, and
Noguchi consulted "specialists and competent clinical observers" who
examined both the monkeys and microscopic sections of their diseased
tissues. Noguchi said that all these men believed that the disease was indis-
tinguishable from trachoma in man, and he told the Board of Scientific
Directors of the institute that he planned to investigate the possibility of
using cultures from the organism for the preparation of a preventive
vaccine and therapeutic serum.[24]

In a paper that he presented at the annual meeting of the American Medical Association in the spring of 1927, Noguchi implied that an organism he had found was the cause of human trachoma. Once again he was prematurely honored, this time by a medal for this work given by this association.[25]

But the news from West Africa had to be faced. Bauer had gone out there in the fall of 1926 and was joined by Stokes from England in May 1927. On June 15, Stokes wrote that they had produced what appeared to be yellow fever in chimpanzees and crown monkeys, but had failed to find *Leptospira* or to transmit the disease to guinea pigs.[26] But the possibility of having a susceptible animal to work with was important progress. Bauer had been keeping Noguchi fairly well informed of all that had happened in Africa before this event, but now he was too busy to write. It annoyed Noguchi that he had to get word second-hand, through reports from the foundation, and he let his annoyance be known. Assurances were given that Bauer would write Noguchi "from time to time," but the only further letter that Bauer wrote Noguchi was not done until the end of September. By that time Bauer knew that Noguchi was expected to leave for Africa shortly, and his letter did not arrive until after Noguchi had left New York.[27]

Miss Tilden recalls that when Noguchi went up to Shandaken for the summer in 1927, he took the "Schüffner papers" with him. Published in German, these papers reported studies by Dutch workers on the three so-called forms of *Leptospira*, a repetition of Battistini's studies but with opposite results. Their findings were essentially the same as those that Theiler and Sellards had reported but extended in these studies to include seven-day fever.[28]

Noguchi also must have taken with him a paper of Sellards' that reported still further studies as though the writer's primary objective had been the annihilation of Noguchi's yellow fever work. Sellards analyzed Noguchi's observations in detail, pointing out the many inconsistencies. He followed with a report of his own trip to Brazil and of elaborate experiments that discredited every detail of Noguchi's work in every possible way. Sellards was meticulous to the extreme of appearing just plain nasty, but he established his point: that Noguchi's serologic work was highly questionable.[29]

Sellards continued his seemingly vengeful pursuit of Noguchi's yellow fever errors still further, publishing still another paper with Theiler in November 1927 that further attacked his prey with the same glee.[30] It has been said that Sellards hated the Japanese, and that he was an excitable man whose reports of events were not to be trusted even though no one questioned his scientific work. He and Theiler had a falling-out shortly thereafter, and Theiler left Harvard. But Theiler would not comment on Sellards' personality or motives, and no reliable explanation has appeared for Sellards' vindictiveness, but for Noguchi such ridicule was intolerable.

Miss Tilden says that when Noguchi returned to the institute in September he had made his decision to go to Africa. He had spent the whole summer at Shandaken winding things up in preparation for his departure as he had never done before. On August 20 he had written a long letter to Flexner. Beginning with a summary of a three-part paper he had written on trachoma, Noguchi remarked with apparently deliberate casualness "Perhaps when an opportune time arrives it may be worth while to make an African trip to study trachoma in Egypt and, if time allows, to visit the yellow fever regions on the West Coast at the same time."

Turning to the verruga-Oroya fever work, Noguchi told Flexner that he had been invited to a conference in Lima in October but did not think he should take the time to go. "It may be, however, of some interest to the members of the conference if we sent some experimental work for exhibit." He outlined the work that he thought appropriate for demonstration and asked:

> Could it be possible that we send some one to represent me and explain the exhibits and read papers? I have in mind Miss Tilden for this mission as she understands everything regarding our verruga work, has followed the experiments from the beginning, and knows Dr. Lorente and many prominent physicians of that country. *Besides, she can bring back to the Institute specimens of insects from Oroya fever regions for determining the actual carrier of this infection.* The solution of carrier problem is not an entomological, but bacteriological one because we now know the method of cultivating the Bartonella from the material suspected to carry it.[31]

With enthusiasm that grew as he wrote, Noguchi explained that material could be gathered for Miss Tilden by someone in Dr. Lorente's laboratory. "She does not have to go to the danger zone at all. . . . I am too busy with trachoma to do this in person, but if we could determine the verruga carrier it will be a great contribution, and I think we can. Of course, I have not spoken of this to anyone. Miss Tilden has given meritorious service in the past 10 years and this little trip abroad may not be an over generous award especially in view of a useful scientific expedition she can fulfill."[32]

If this was only the germ of an idea when Noguchi started this letter, he became so pleased with it that he added a long postscript in which he suggested that exhibits be sent to show the trachoma work as well. Noguchi said that he would prepare three papers on verruga and one on trachoma, but as he listed these papers he added still another. Viewing this hyperactive planning as a symptom of his mounting anxiety, is it not appropriate to wonder whether this proposed trip for Miss Tilden was not his first farewell gesture, a gift to her for her years of devotion? Noguchi apparently liked the idea of winding up the verruga-Oroya fever problem this way. Nevertheless, the activity and apprehension that developed as he declared his intention of returning to yellow fever work pushed all consideration of this suggestion aside after his return to the institute in September.

Okumura writes that Noguchi's neighbors at Shandaken found him looking older that summer of 1927. They often saw him sitting on the porch of his house staring off into the mountains. Although he had been pushing Flexner to let him go to Africa, Noguchi touched on the matter only lightly as he suggested combining it with a trip to Egypt to study trachoma. He had already decided to bypass Flexner and to notify the foundation directly that he was going to Africa.

When he returned to New York, Noguchi sent a short, typewritten note to Dr. Frederick Russell at the foundation in which he discussed another matter, but to which he added a handwritten postscript: "I would like to see you sometime before very long. H. N."[33] Only a week later, on September 14, Russell sent a cable to Dr. Henry Beeuwkes, who was in charge of the Rockefeller Foundation headquarters at Lagos, saying that Noguchi could leave for Africa in the near future.[34] In addition to news that Stokes and Bauer had transmitted a disease that was thought to be yellow fever to monkeys, Noguchi's commitment was now deepened by another bulletin — that Stokes had contracted yellow fever.

It is commonly assumed that Stokes's death from yellow fever was the final stimulus for Noguchi's decision to go to Africa, but this cannot be true. Stokes was first hospitalized on the morning of September 15, the day after Russell's cable was sent. But the fatal outcome of Stoke's infection certainly precluded any further objections from Flexner. Stokes had requested that his blood be drawn for injection into experimental animals, and that mosquitoes be allowed to feed on him during the acute stage of his illness. After the necessary incubation period, the mosquitoes were placed on monkeys and there was no doubt that the disease was transmitted. Stokes died on September 19, and autopsy findings demonstrated conclusively that his illness was yellow fever. No *Leptospira* were found.[35]

Now the preparations for Noguchi's expedition became very active, both in West Africa and in New York. Russell wrote Beeuwkes that he had had a "very satisfactory interview with Doctor Simon Flexner and Doctor Noguchi yesterday," and he gave Beeuwkes a long list of instructions. Financial considerations came first: Beeuwkes was not required to provide for Noguchi's needs out of his budget, but separate funds would be made available from New York. "I feel that it is great good luck that Doctor Noguchi is considering and will probably go to West Africa and that there is nothing too much that we can do to make his visit pleasant, agreeable and profitable. He is, as you know, a unique personality and an indefatigable worker, who has his own methods, his own techniques and the imagination of the real research man." The staff members at Lagos, where the foundation facilities were located, were to turn over everything they had to Noguchi. If his needs for a laboratory and for animals should interfere with work that was going on, additional buildings should be constructed. Since Noguchi was thinking of starting his work at Accra on the Gold Coast (where cases of yellow fever were more plentiful) Russell suggested building an animal house there as well. Living quarters should be arranged for Noguchi at Accra as well as at Lagos.[36]

Russell said that Noguchi did not wish to have his presence interfere with Dr. Bauer's taking the leave that was now due him, and then Russell added a paragraph which also appears in Flexner's papers without identification. It can be recognized as an addition to Russell's letter written for the purpose by Flexner:

> At the risk of repetition I might say in brief that Dr. Noguchi is the type of intensive individual worker who knows precisely what he needs or thinks necessary and essential. He is a courteous, delightful gentleman, and of course in laboratory experience far beyond anyone else who has been with the expedition. Although you do not need to be told it, yet it can do no harm to say that he is to be his own master and is to have all the facilities and provisions for his work he asks for, and given without delay. He will make no unjust demands, and yet he may ask for things which you may not think necessary. But as Dr. Noguchi is the distinguished guest of the Foundation and a most able and eminent worker, there is nothing that he can ask for that should not be granted at once.[37]

Kligler, Noguchi's co-worker in Mexico and Peru, could have identified the writer of these words in an instant. Did Noguchi, again, ask for such an introduction? Flexner wrote later in his diary of these last days:

> His going to Africa was a great hazard. We all felt this. He had wished to go and I obstructed year by year. I feared his health; it has not been overrobust. He has got stronger. . . .At last, I gave consent provided his doctors passed him. Cohn examined him and I think Libman also. He was passed. Then I arranged for the bust by Konenkow. This was carried through with Levene as intermediary so that he should not suspect the fear behind it. N. was delighted to do it. He took it as an honor; it was his way. He even directed K in the posing, and would be modelled looking serious—as a Samurai— as I thought. The results pleased him very much. N. was always sensitive of attention.[38]

On October 13, Flexner wrote Noguchi: "It was a great pleasure to have you dine and spend the evening with us." He enclosed letters from Africa that reported the most recent findings, assuming that Noguchi had already seen them but adding his comments:

> There is one thing I think we should keep in mind. It is really not necessary to speak of it, but we want to do everything for comfort

and in proper order. If by any chance Stokes and Bauer did communicate some disease from cases to monkeys, the discovery is of course theirs, although you may be able to extend and define it. You will of course be of the same mind about this point as I am. I think it would be well for you (or me) to say just a word to Col. Russell about the matter, unless you have already done so.[39]

There were undoubtedly those who might have said that Noguchi deserved this, perhaps Battistini among them. Whether it was justified or not, only five days before his scheduled departure for an expedition that both Flexner and Noguchi feared might put an end to their intimate association of twenty-seven years, Noguchi had to stop and reassure Flexner. Noguchi's anger was not concealed by the formality and courtesy he could not abandon even in such circumstances. As it is clear that Miss Tilden edited his reply, it is equally clear that she shared his outrage:

It was very kind of you to have me dine quietly with Mrs. Flexner and yourself the other evening. I thoroughly enjoyed being taken back to the sweet old days that are passed but are still vivid in memory. Thank you very much, too, for the copy of Claude Bernard's *Experimental Medicine,* which I am happy to have.

I have your letter of October 13 with regard to the work carried out by Stokes and Bauer. It is not altogether clear to me what you mean, but I take it that you are advising me to give them a square deal and not to appropriate their findings. . . .Your apprehension should be relieved, however, when you learn that I have already spoken with Dr. Russell. . .asking that the data obtained by Stokes and Bauer be gathered and given immediate publication so as to insure the priority of their work. . . .I want to assure you again that Stokes and Bauer will not be robbed of their accomplishment because of my going out there. . . .I hope that I have made my position clear to you before I embark on this expedition. I should be glad if you will show this letter to Dr. Russell.[40]

Noguchi had never been accused of appropriating the work of others before Battistini's rumored accusation; this was not his way. It must have been clear to Noguchi, as it had been to others for a long time, that Flexner's first consideration in any situation was the interests of the institute and the foundation, as he saw them. No other person or thing could ever compete with Flexner's primary objective—the promotion of the good name of these institutions. If the institute could be seen to suffer from Battistini's allegations, whether or not they were justified, then Battistini was not a reliable man and should be allowed to sever his connection with the institute just as de Kruif had been encouraged to do. If, after all the years of their association, sending Noguchi off with this insulting directive might further undermine his already wavering stability, so be it; the reputation of the institute for doing "everything for comfort and in proper order" took precedence over any one man's survival.

According to Miss Tilden, Noguchi gave her a few hundred dollars and asked her to keep them for him. In case he should not return, he wanted her to send the money to his sister in Japan, giving Miss Tilden the impression that he did not trust his wife to do this. Miss Tilden followed his instructions, even finding someone whom she trusted to carry the money rather than sending it through the mail. She seemed to be unaware that Mrs. Noguchi had been given her own set of instructions and had carried them out with the same meticulous care.[41]

Okumura writes that on the evening of October 19, Noguchi went to a party given by some Japanese friends, but he was very quiet and left early; he went back to his laboratory and did not get home until four in the morning. The next night he stayed all night in the laboratory, returning home at seven in the morning. He slept a little, and when he awoke he prepared to leave for the ship. Included in his luggage was a paper scroll, a tracing of his mother's death name taken from her tomb. His wife, already frightened, begged him to leave the scroll but he would not listen. He did leave the talisman that had been sent him every year from the temple, first by his mother and after her death by Kobayashi. He had taken one with him on all of his previous expeditions, but this time he had left it in the laboratory.

It would be easy to dismiss this sentimentalized foreboding as having grown in the hindsight of those who wrote of it.[42] More difficult to dismiss, however, is the fact that almost all persons who were available to talk about these events, spontaneously brought up their recollections of the staff meeting that was held the day before Noguchi's departure. Dr. Alexandre Rothen, who was new at the institute at the time, and who knew Noguchi and Miss Tilden only by sight and reputation, has a vivid recollection of riding with them in the elevator; no words were spoken, but as they stood together with their eyes straight ahead, she so tall and strong-looking and he so little, they created a somber mood that Rothen has never forgotten. He followed this story with the remark that there was a general feeling at the institute that Flexner was very hard on Noguchi.[43]

Dr. Doan wrote of Noguchi's talking about yellow fever, saying he would be successful in his mission and would come back with the answer.[44] Dr. Philip McMaster said that it was his responsibility at the time to arrange the programs for these staff meetings, and he asked Noguchi whether he would like to report on his trachoma work. Noguchi agreed and did so much as any staff member would do at any meeting.[45] Consistent with McMaster's recollection is that of Dr. Rebecca Lancefield: she remembers Noguchi talking about trachoma. Nothing unusual was said, neither Africa nor yellow fever was mentioned, but she found it one of the most moving experiences of her life. It seemed to her that they all suddenly became aware that this was a man whom they did not know at all in spite of their years of association; that behind his child-like charm, his courtesy and arrogance, he was a man who was subject to the same emotions as everyone else.[46] They were startled to realize how unaware of this fact they had been, and they now recognized the courage with which he was facing the real danger as well as the humiliation that lay before him. They all believed that he had been mistaken about yellow fever, and that he knew it at the time of his departure. McMaster said Noguchi knew he had been wrong before he left and perhaps it would have been worse for him if he had lived.

Noguchi sailed for Africa on October 22, 1927. He wanted no one to see him off. Miss Tilden had wanted to go to Africa with him; she feels that things might have turned out differently if she had gone, but Noguchi would not allow it. Philip Reichert says Noguchi had asked him to go and he wanted to do so, but had to refuse because he was the sole support of his mother and younger sisters.[47] There were only Noguchi's technicians at the pier to help with his baggage. We are told that Noguchi begged his wife not to cry but to "help me go."

Flexner attended the final staff meeting and left for Europe shortly thereafter. He returned to New York for Christmas, and on January 3 he wrote Noguchi in Africa. He thanked Noguchi for his Christmas greetings, and told him to take care of his health, to do this and that in his yellow fever work, and then:

> Well, I wish you in this new year not only preservation of your health, but success in your work and especially happiness in your life. I am enclosing an intimate account of you which appeared in the *New York World* yesterday. Some of your Japanese friends doubtless were called on for the material. It is a very sympathetic story and one that can only give pleasure and do good.
>
> In reading it through, I was impressed with my own lack of knowledge of your early personal history. I wish some time, merely as recreation, you would write down something of your boyhood, your parents, early life, education, ambitions, etc., up to the time you arrived in Philadelphia. I am confident that the account would be very interesting. I do not mean you to write it for publication, but for my information and for the records of the Institute. We have quite a few things collected and put away in a safe place, and I think it would be very valuable to have a reliable record of your early history. . . .[48]

22

West Africa

Dr. Noguchi was not himself in West Africa.

—Evelyn B. Tilden[1]

DURING THE SIX MONTHS KLIGLER SPENT SEARCHING FOR *LEPTOSPIRA* in yellow fever cases in West Africa in 1926, he wrote his wife: "I shall have some interesting tales to tell you about the Commission of ours but these things cannot be put in writing. You will have to wait until I return. Human beings and human frailties and human jealousies and human bigotry are the same the world over and they are as humorous here as they are elsewhere to an observer who can stand aside and see the comic aspect of it all."[2] Shortly after the completion of his assignment for this Rockefeller Foundation Expedition Kligler wrote to his wife: "But something quite revolutionary may develop from our work. Evidence is accumulating that the causative organism of this disease is not the same as that in South America. If that proves to be the case it will be one of the big jokes of the day."[3] But it was a rare man who could find humor in the traps set by yellow fever or in the foibles of the people who were caught in such traps.

Nigeria and the Gold Coast, just above the equator, were both protectorates under the British Colonial Service but separated from each other by German Togoland and French Dahomey. The Colonial Service maintained a hospital and laboratory at Accra, on the Gold Coast, whereas the yellow fever headquarters of the Rockefeller Foundation expedition were farther south at Lagos, Nigeria, an overnight journey by ship. Overland travel between the two areas was limited by poor roads and the lack of railroads across the French and German territories, and telephone service did not exist. Coastal ships carried mail as well as passengers; communication was otherwise carried on by means of cabled messages. Although Nigeria was often called the white man's graveyard because of the prevalence of tropical disease, yellow fever cases were reported more frequently at this time from the area around Accra, perhaps in part because of the trust inspired by Dr. Alexander Mahaffy, an International Health Board staff man who was permanently assigned to Accra. The local inhabitants were reluctant to report their illnesses to foreign medical doctors, but Mahaffy often established confidence where others failed.

Conditions were trying for workers in these isolated outposts, even though the climate, unlike in many tropical areas, was relatively benign. Living quarters, too, were adequate and comfortable. The real danger of infectious disease, of which the mosquitoes were a constant reminder, and the enforced companionship with a small group of colleagues exaggerated and compounded the personal problems each man brought with him,

Hideyo Noguchi and Alexander Mahaffy, Accra, 1928. The Rockefeller University Archives

lems that were often the reason for his having accepted assignment to such a post.[4] Even though there had been considerable progress toward the revolutionary development foreseen by Kligler, one member of the team had died and there was no levity at Lagos. Neither was the atmosphere brightened by the elaborate preparations for Noguchi's arrival or by the orders to grant his every wish without delay.

Noguchi sailed for Liverpool on October 22, 1927, and spent much of his sea voyage sleeping. The anxiety that had pervaded the weeks of decision and preparation for his trip had been so exhausting that it had overwhelmed his habits of a lifetime. When he arrived in England Noguchi called on professional friends who were acquainted with medical problems in West Africa, and he started an active response to cables from Miss Tilden and Beeuwkes, a response that kept him busy all the way to Accra. From Miss Tilden he learned that Bauer's letter had arrived with the promise that citrated blood from old cases of yellow fever, as well as a tube of Stokes's blood, were being saved for him. Beeuwkes reported that there were no new cases of yellow fever at the moment and Noguchi later wrote Flexner: "In contradiction to Dr. Bauer's letter, Dr. Beeuwkes cabled to my boat that no blood was saved from Stokes's case and that all specimens were about four months old."[5] Noguchi was already looking for trouble and started taking protective action even before he landed.[6] For the entire six months of his stay in West Africa, Noguchi kept up a feud with the people at Lagos, and with Beeuwkes in particular. Although the group at Lagos tried to assure Noguchi and the New York office that they wished to give him their full cooperation, it was beyond their capacity to do so, partly because of Noguchi's truly excessive demands and suspicions, and partly because their jobs were difficult and worrisome enough without the extra burdens put upon them by Noguchi. Beeuwkes was a retired army officer who was so meticulous that, according to one of the junior staff men, he required all personnel to turn in pencil stubs before they could be issued new pencils.

Throughout his stay in West Africa, Noguchi's cabled messages always expressed satisfaction with his working conditions even when his many lengthy letters were reporting all kinds of problems and persecutions. He took over the ordering of experimental animals from European dealers because he believed that Beeuwkes was deliberately keeping him in short supply. Although Beeuwkes reported that he had ordered enough to ensure "well over 200 monkeys arriving within six weeks from the time Dr. Noguchi reached the coast,"[7] it was not enough for Noguchi who, at one time, had some 400 monkeys under observation.[8]

Noguchi's first letter from Africa to Flexner, dated January 19, 1928, was twenty pages long. Noguchi wrote that blood samples had arrived from Dakar and reported that he had found them infective and had started a strain in animals. When no more fresh animals could be expected, Noguchi went to a local village himself to track down a rumored case. Shortly after his return he fell ill with a malady, which added to the general confusion. At the hospital at Accra Noguchi's case was treated as dysentery since the only observable symptoms fit that diagnosis. He wrote to Flexner, however, that his illness had been a mild attack of yellow fever which none of the doctors had recognized. Noguchi had asked his technician to take some of his blood and inoculate a monkey, and he enclosed a picture of a sick monkey to substantiate his report of the outcome:

That rhesus which was inoculated with my blood on January 2nd came down with black vomit and slight jaundice on January 17th....It was one of the most typical pictures I or anyone else had ever seen. I autopsied it within 30 minutes after death. . . .Of course all physicians and Commission members who saw my case and failed to recognize the nature of the disease were completely dumbfounded. But, I knew from the beginning I will get over without much consequence because I have been vaccinated just before starting on the present expedition. No matter what anyone says the icteroides vaccine protects against yellow fever, completely or partially. . . .I am working with the Dakar strain and now will have a "Noguchi strain"! in addition. . . .I am now immune both to the American and African yellow fever and can safely carry out experiments without further accidents. What an odd incidence! Stokes died, he did not believe in icteroides vaccination and had none — and he died.[9]

Dr. Mahaffy made a summary of the medical records of Noguchi's illness, noting none of the yellow fever symptoms that Noguchi reported, none of the commonly recognized symptoms of yellow fever.[10] The mystery of the apparent communication of the disease to the monkey was not cleared up for the people in New York for some months, but those at Lagos were quite sure they knew what had happened.

Noguchi's monkeys were kept two in a cage, and, on occasion, their tags would slip off. The boys who cared for the animals had been told that in such an event they were to bring both the animals and the tags to Noguchi, but they often just put the tags back on and said nothing about the matter. A young medical student, A. P. Batchelder, had been caring for mosquitoes, which were being maintained for experiments, until he was fired when Noguchi claimed that he had killed them all off by improper feeding. After Batchelder returned to New York in August 1928, he told Muller of these customs of the laboratory boys as well as of other evidence that the monkey that had been inoculated with Noguchi's blood was not the one which came down with yellow fever. Batchelder described some immunity tests that had been performed with serum from various people, both natives and foreigners. Some were known to have had yellow fever, the others were known not to have had it. Noguchi's own serum was found to react in the same way as the nonimmunes. Muller reported Batchelder's observations in a letter to Flexner, stating in summary, "this inclines me to think that the virus called the 'Noguchi strain' was accidentally derived from some other source."[11]

In other words, the condition of Noguchi's laboratory was chaotic, because of overcrowding and his old preference for using untrained helpers, as well as his inability to supervise under the best of conditions. Noguchi worked mostly at night and thus had limited opportunity to watch or speak with his helpers, and he worked such long hours that they hesitated to call upon him for clarification when they had questions. Yet Noguchi continued to insist that he was doing constructive work (some of it was) and that he was making progress even though his desperation was clear in all of his letters. To Flexner, in the long January letter, he wrote: ". . .during the rather brief period I have learnt much — gone through a natural infection and getting quite near the solution of the vexed problem. I am indescribably happy that you have so loyally and kindly backed me up when I was being mercilessly ridiculed and pressed for action. I owe you this all once again. . . .Please have faith in me, I will bring back something worthy of your unqualified trust."[12]

Russell had countermanded Noguchi's authority to order animals from the European dealers, promising a letter that would explain the necessity for this action. When Noguchi received the explanation — that Russell considered it important for the future of the program to maintain continuity — he wrote on the margin of Russell's letter: "I wonder what he actually means by future. I am going back soon and no one will need animals here after my departure. Never mind ten years! H. N."[13] Noguchi dispatched the letter with a surly note to Flexner reporting that he had notified Beeuwkes that he no longer needed Batchelder or another Lagos man who had been helping him.[14]

In marked contrast to the many long letters he wrote to other people during these months, Noguchi's brief letters to his wife made no mention of any special problems. He was consistently optimistic about the results of his work, reassuring about his health, and pleased with his surroundings and the cooperation he was being given. One such letter was dated "Xmas Eve. December 24, 1927."

My darling Maizie:
. . .I am hoping soon to discover the germ and find a method to protect people from yellow fever. It is not the same disease that I had studied in South America.
As I wrote you before I am well fixed and my health is excellent. This is the Xmas eve and I miss you and I know you miss me, too, but as you know I must find the germ. I hope you are doing alright. Don't worry about me. Read good books, go out to movies and shows and don't get lonesome. I will be back as soon as I find the germ. Is Tom coming often to see you? Give my regards to him and Jack and Andy. I have no time to write to any one even to Dr. Flexner. This evening all American doctors here are going to have supper together and I will have to be with them. They are all working for me.

Hideyo Noguchi and Johannes Bauer, Accra, 1928. The Rockefeller University Archives

> Now, I must go back to the laboratory so will close with love.
>
> From your own Hide.[15]

By the time of Noguchi's January 19 letter to Flexner, it can be seen that his suspicions were being encouraged by Bauer, who was currying Noguchi's favor in order to enlist his help in getting a job at the institute. Noguchi spoke of Bauer at length:

> He wishes to work under me when I return. I like him very much. . . .He does not care to come back to Africa. Can you give him a place at the Institute? He is receiving $300 a month, far less than other clinical members of the Commission. Of course, yellow fever work is or will be much cleared up before I leave here. He has been working under trying conditions and does not wish to renew his appointment under the present director. Besides, from my own experience, there is a great danger to his life while working with yellow fever.[16]

In a postscript Noguchi told of the work Bauer had been doing, mostly with mosquito transmission, and continued:

. . .nevertheless, he is exposed to laboratory infection. His term expires
on February 26 and he wants to wait until my work is finished so as to go
back together. But, now I do not see how I can finish my work before
the middle of April. I really do not want him to stay any longer than absolutely necessary for him and I cannot give or assign him any part of
my work. He is, I believe, best recalled or ordered to take his coming
leave as soon as his present term expires. . . . There is constant danger
. . . . I do not want Bauer to stay in Africa a day longer than his term
(February 26) as I cannot be responsible for any accident. Dr. Russell
knows my attitude on this matter.[17]

Bauer did some of the best work that was done in these investigations,
and may have feared that the major credit would go to Stokes who was
already well-known in the field and whose death had made him a martyr to
the cause. For whatever reasons, there is no question that Bauer took advantage of Noguchi's disturbed condition to advance his own position.
Evidence that there was more to this than Noguchi's agitated distortion is
found in two letters from Bauer. One came in the middle of March and
Noguchi saved it; but when he received a similar letter a week later, he sent
them both to Miss Tilden, asking her to keep them until he could return
and discuss them with Flexner.[18] Bauer wrote:

Many thanks for your very nice cable and a letter which I received last
mail. As a matter of fact, I was so glad to receive your letter, especially I
had not heard from you for a long time, and I thought you were sore at
me for one reason or another. I was extremely sorry to learn, however,
that you are not feeling at your very best. And no wonder, I think that
all the obstacles that were put in your way here would break any man
down. I think that everything was done here that was possible to make it
more difficult for you. . . . You know I have finished my 18 months
already, and I don't remember of having been in any place I hated more
than I hate this place here. It is not the climate, and I am not tired, but
it is the atmosphere of spying and all sorts of dirty intrigues that is most
distressing. I am absolutely earnest when I tell you that I shall be you
most grateful for the rest of my life, if you will help me somehow to get
me some kind of job at the Institute so that I don't have to come back
here again. At times it seems to me that I would take even a janitor's job
in order not to come back for another tour.[19]

Bauer's second letter dealt more directly with their work but in the same
critical vein.[20] By this time Noguchi had made an apparently independent
observation that added to his legitimate concern. Before Stokes's illness it
had been the general opinion that there was no possibility of exposure
other than through the direct inoculation of the organism into the
bloodstream; in practice this could be accomplished only by the bite of a
mosquito. After Stokes became ill, however, his co-workers at Lagos "conducted a series of experiments in monkeys. They showed the ease of infection through the skin from materials being routinely handled. These findings caused us to increase our precautions above the usual measures we had
used."[21] Whether or not Noguchi knew of these experiments, he wrote

Flexner: "Contrary to the general dogmatic statements I found that the blood and various organs removed at autopsies (monkeys) are highly infectious."[22] But Noguchi could add little to his precautions. He performed autopsies without gloves since he could not wear a glove on his maimed left hand. Was he thinking of his own danger when he wrote so much of the danger to Bauer? Was his insistence that his illness had been yellow fever, and that he was now immune, a psychological defense against the fear of his true vulnerablity?

In spite of denials and in spite of a general recognition of the special reasons for Noguchi's anxiety, the workers at Lagos were very uneasy; they knew of his fear of humiliation at being proved wrong as well as the physical dangers. Noguchi's behavior and arrogance certainly added as much to their uneasiness as their manifestations of anxiety added to his. But his relations with the people at Accra were altogether different.

When Noguchi first arrived in Accra, he found that arrangements had been made for him to live with Dr. and Mrs. Mahaffy because their house, although three miles from the laboratory, was in an area relatively free from mosquitoes. Within a few days Noguchi discovered that the medical student Batchelder had a bungalow "all for himself" and Noguchi asked that a bungalow be screened for his use. This was done and he lived there until he entered the hospital. After his illness Noguchi moved back to live with the Mahaffys "for a better after-care and also to prevent my overwork." Bauer had come up from Lagos to spend a few days with Noguchi right after his arrival to bring him up to date, but otherwise Noguchi's personal dealings had all been with people associated with the British Hospital at Accra, including Dr. and Mrs. Mahaffy.

After Noguchi's recovery from his illness, Dr. William A. Young, a pathologist who was director of the British Hospital, offered to help him with his work. Noguchi accepted this offer, and Mahaffy reported later that Young thereafter gave most of his time to Noguchi's projects. Young participated in tissue work: ". . .he has been cutting and staining sections for me. . .it was agreed from the beginning that he was to give me the pathological diagnosis of the tissues without any knowledge of the source of the materials, so as to obtain as strictly objective observations as possible. It was through such an arrangement that we became more intimately acquainted with each other."[23]

In February, Noguchi wrote both Flexner and Miss Tilden (finishing lengthy letters at 5:30 A.M.) that he had five yellow fever strains, including the Noguchi strain. He told Flexner that yellow fever in Africa did not conform to the classic description of the disease by the Reed Commission. Noguchi claimed that he had discovered from his own material that the incubation period could be as long as sixteen days in monkeys, an observation made independently, and later published, by Bauer.[24] It had been assumed that the accepted incubation period for yellow fever in man of

three to six days would also apply to monkeys. Noguchi also claimed that
he had found an organism that produced lesions like yellow fever. He was
aware that his organism might be a contaminant but felt he should follow
it up.

On March 19, Noguchi wrote Flexner: "After all I got the organism that
causes yellow fever in Africa." Two monkeys that he had inoculated with
"pure cultures" of his organism derived from old, preserved specimens of
yellow fever blood, had come down with the disease and had died during
the previous night. Noguchi had been so sure of what the findings on
autopsy would be that he called Young out of bed to come as a witness.
When Young declared that the cases were yellow fever and learned the
source of the strain with which Noguchi believed they had been infected,
once again he was "dumbfounded."[25]

Now Noguchi started a series of experiments to demonstrate for Young
and Mahaffy the whole process through which he believed he had isolated
and cultured the organism and produced true yellow fever in animals. He
would take them through his work step by step, giving them full opportunity to point out any error he might have made. Confident that they would
observe what he had reported to Flexner, Noguchi wrote: "I have never
been so certain of being correct as I am with this piece of work on West
African yellow fever. God bless you for your big heart that protected and
guided me through many trying years I went through since my Guayaquil
expedition. I am a better and sounder worker now than I have ever been,
so please do not worry."[26]

Noguchi wrote essentially the same thing to Russell, describing his past
relationship with Young: "It was agreed that he shall not be informed of
my work until completed because of my peculiarity in solving any problem.
. . .It is fortunate that Dr. Young is one of the best yellow fever experts on
the West Coast and can intelligently and critically check up on my findings. . . .These findings. . .have been kept to myself because I do not
believe in communicating every bit of new finding to another man who is
interested in the same subject at the same time. It only disturbs his in-
itiative and helps nothing. . . .When Bauer comes here on April 21 he will
be shown all, but not until then. I hope Dr. Beeuwkes will not misconstrue
my silence as sign of discourtesy." He added his feelings of gratitude to
both Dr. and Mrs. Mahaffy, congratulating Russell on having "this type of
man on your staff out here. The British like him. He is a good clinician as
well."[27]

We cannot tell just what Young saw or even whether Noguchi completed
his so-called demonstration. Noguchi intended to withhold nothing and
wrote Flexner that unless he could convince Young the work would not be
much good anyway. Neither can we tell whether Young was convinced of
anything but Noguchi's profound conviction. He knew that the laboratory
animals had died of yellow fever, but we cannot tell how closely he was able

to follow the experiments. Noguchi had heated his cultures to 80° C for fifteen minutes, which he was sure would destroy all possible contaminants. When the residual material produced yellow fever, Noguchi concluded that the only organism that could have survived must be the yellow fever organism,[28] a dubious conclusion at best for a scientist, and more than dubious if the conditions of his laboratory were taken into consideration. Theiler later reported that dried yellow fever virus does survive not only freezing but also such prolonged high temperatures as Noguchi had reported, so the critics who cite this claim as evidence of his incompetence are not justified in doing so.[29] But no one was able to find the organism that Noguchi identified, either from his material or from attempts by others to repeat Noguchi's experiment. Noguchi had described an "ordinary-looking" organism that he said others would likely ignore and that was protected from the high temperature in spores which it produced.[30] Once again, Noguchi made the mistake of assuming that if the organism could not be visualized with his technique, it did not exist; and conversely, any organism that survived his own method of isolating it from contaminants must be the causative agent of the disease. He knew better than this.

On April 21, Bauer and Hudson came up from Lagos. Bauer remained one day and then continued on his homeward journey, but Hudson spent five days going over Noguchi's material before returning to Lagos. Bauer reported in New York that Noguchi had seemed confused on that day, but we know no more of his visit than that. Paul Hudson was the pathologist who had arived in Lagos in August, just in time to participate in the successful transmission of yellow fever to monkeys. He had confirmed the diagnoses in the animals through his examination of their tissues at autopsy and was the third member of the triumvirate—Stokes, Bauer, and Hudson—who published the reports of these first successful transmissions.[31] Hudson had been in Lagos long enough to become friendly with Adrian Stokes and, as pathologist, he had to perform the autopsy that confirmed the cause of Stokes's death.[32]

Noguchi was so mistrustful that he sent a cable asking Flexner whether he should give samples of his cultures to Hudson.[33] Noguchi had been working at a pace that was excessive even by his standards, and the strain was showing in his appearance. A picture of Noguchi and Hudson, with a microscope between them, shows an aging Noguchi, drooping, dead-eyed—and for the first time, with his maimed hand in full view of the camera. He gave Hudson tissues from seven monkeys, asking for a full report. But, just as he had done with Young, Noguchi gave no further information. Hudson's report was not transmitted until after Noguchi's death, and his covering letter was guarded. Hudson had probably heard of the mess in Noguchi's animal house and would not even identify sections with yellow fever because he was "not cognizant of the whole experimental history of these animals."[34]

On the evening of May 4, Noguchi called to say good-bye to Young's wife who had been at Accra during the whole of Noguchi's stay there and who was leaving to visit her home in Scotland. Touched by this visit, Mrs. Young later wrote:

> My husband and Dr. Noguchi were very great friends and worked so hard, by day and at night. . . .The evening before I sailed from Accra, Dr. Noguchi came round to say good-bye, and told me so many times how greatly he appreciated all my husband had done for him, he thanked me time after time. I think he felt he could not thank him enough, and so wanted to thank him through me. I told my husband so, and he smiled and said he knew."[35]

Beeuwkes came through Accra on his way back from a yellow fever conference at Dakar, and on May 9 Noguchi joined him on a ship bound for

Hideyo Noguchi and Paul Hudson, Accra, 1928. The Rockefeller University Archives

Lagos. Noguchi had never visited the group at Lagos in the five months he had spent in Africa. He was very tired, but still wanted to tell Beeuwkes all that he had accomplished. By the time Beeuwkes got to his diary, even he was impressed and dizzy. "It all sounds most remarkable and interesting, and I trust that the results will be more or less epoch making. It would be presumptuous for me to attempt to outline his work or evaluate it. . . .He is not yet certain himself where he stands and will carry a large amount of material to New York for continued study. . . .His findings are less uniform than in experiments carried out at Lagos, and this may be accounted for by the virulence of the strains. . . .The point that impressed me most was the enormous field that Dr. Noguchi has covered in such a short period of time."[36]

At Lagos, Noguchi spent all day Thursday and Friday morning in a busy round of activity. He had a slight chill Friday morning and asked Hudson to examine his blood for malaria. The smear was negative and his temperature was normal at that time. When he boarded the ship around noon, Noguchi spoke of being very tired and went to bed; at Accra the next morning Mahaffy found him very ill.

After a difficult landing through the surf, Mahaffy took Noguchi back to the Mahaffy home and saw that he went to bed. By late afternoon his temperature was 103° F; he had a headache and other symptoms suggestive of yellow fever, and he was hospitalized that evening. Mahaffy's report of the week that followed is professional, but far from detached:

Dr. Noguchi realized Sunday morning that he was suffering from yellow fever. On Tuesday he appeared better and this apparent improvement continued on Wednesday. Wednesday morning he asked for the electric kettle and tea tray, which was always kept in his room here, in order that he might make his own tea. On Wednesday morning I read to him the messages which had been received from New York and he seemed very grateful. . . .When I saw him at 6 a.m. on Thursday he did not seem to be so well, and complained of pain in the hepatic area. Failure of liver function was feared but his condition was better in the evening, and on Friday he seemed to be considerably improved. His mind was clear and active Friday afternoon, and he asked many questions concerning the laboratory. He also stated very definitely that he had no idea how he had become infected, and added "We know very little about Yellow Fever." He asked where Dr. Bauer was and wondered if he had arrived in New York.

The epileptiform seizure on Saturday morning came as a great shock to all of us as we had felt that he was practically out of danger the previous day. After the seizure he became drowsy but did not absolutely lose consciousness until Monday morning a few hours before his death.[37]

During the time of Noguchi's illness, Young and Mahaffy burned all of the dead animals that were being saved for Noguchi's return from Lagos. But Young selected a few that he thought were most important to demonstrate the outcome of Noguchi's experiments, and he performed autopsies and recorded his findings before disposing of these few animals. Young also drew samples of Noguchi's blood several times during his illness and inoculated monkeys with it, and later transferred the strain to a second generation. He was deeply concerned to preserve every bit of evidence of the work Noguchi had done.

For most of the forty-eight hours following the epileptoform seizure, Noguchi remained in a semiconscious state; once he awoke enough to drink a large amount of cold water, but several times he lapsed into delirium. His colleagues maintained a constant vigil, treating each symptom of his physical deterioration as it appeared, but knowing that there was little they could do to affect the final outcome. If Noguchi spoke at all, his words were not recorded and he died at noon on Monday, May 21, 1928.

Young, as pathologist, knew what he had to do; barely two hours after Noguchi's death, he performed the autopsy, assisted by Dr. Russell (of the local Medical Research Institute) and in the presence of Dr. Franklin, from Accra, and Drs. Beeuwkes, Walcott and Scannell and Mr. Batchelder, all from Lagos. Young confirmed the diagnosis of yellow fever, and he surely must have observed that the changes he saw in the heart were characteristic of, and unique to, syphilitic heart disease. He must have appreciated further that all the witnesses to the autopsy could draw the same conclusion.

Some four hours later a cable from Russell in New York arrived with a request that either Noguchi's heart and aorta be preserved and sent to New York or that a pathologic description be sent in its place.[38] Young did his part in complying with both of these requests. He described the condition of the heart in his report of postmorterm findings and, according to Beeuwkes: "Dr. Young had preserved the more interesting part of the organ [the heart] and he promises to turn this over to me for shipment to New York."[39] Young had already been moved by the intensity he had seen in Noguchi, and he now could recognize that it represented this additional burden.

Eight days after Noguchi's death, Young died of yellow fever. During his illness Young, like Noguchi, said he had no idea how he had been infected and, like Noguchi, showed no evidence that he had been bitten by a mosquito.[40] It could only be assumed that, in spite of precautions, Young had been infected while handling the infectious tissues in his efforts to salvage what he could of Noguchi's work. Young could no more sort out the constructive work that Noguchi had done in Africa from the chaos in the laboratory than he could spare his friend the final humiliation from exposure of the evidence of his old syphilitic infection with all of its degrading implications.

Postscript

So ends in a tragic way and thousands of miles away a life and
an era. But for the moment there was much to do. A press announcement, Mrs. Noguchi to inform, telephones continually.
With Miss Tilden's help I got a press notice prepared. . .to be
given out between 4 and 5 just to escape afternoon papers. . . .
A nightmarish day, of course, and a haunted night.
                                        —Simon Flexner, Diary[1]

THUS DID FLEXNER BEGIN TO ABSORB THE IMPLICATIONS OF NOGUCHI'S
death. People who had been associated with Noguchi in his lifetime now
responded to his death according to their individual needs. Those who had
used him in life continued to do so after his death, whereas those who had
given something of themselves to their relationship now tried to carry out
his wishes.

The institute purchased a burial site for Noguchi, with one beside it for
his wife, at Woodlawn Cemetery in The Bronx. Funeral and memorial services in New York were by invitation and were designed to reflect well on
the institute and the foundation.[2] In Japan, deans of Imperial University
schools of medicine and other members of the academic, political, and
social elite were present at similar services. Even Princess Chichibu, a
daughter of the Matsudaira pre-Restoration Lord of Aizu, attended to
represent the Imperial Household and, symbolically, the Aizu people.[3]

Kobayashi and other members of *Chikubakai,* the organization of
Noguchi's childhood friends, formulated plans for a permanent memorial.
They would restore the Noguchi family home and would build a library
where his publications, letters, and other memorabilia could be collected
and displayed. Kobayashi requested financial assistance from the
Rockefellers for these projects, and further suggested that the Rockefeller

family provide an endowment for a Noguchi prize similar to the Nobel Prize.[4] In the fall of 1928, Kobayashi sent copies of resolutions that had been adopted by *Chikubakai*, together with his requests, and Flexner had these Japanese communications translated and sent to Rockefeller along with explanatory letters of his own.[5]

Jerome Greene, the Rockefeller representative, conferred with these Aizu people when he went to Japan late in 1929. He brought back the recommendation of his Tokyo advisors that any funds should be administered by a Japanese trust company, such as Mitsui, rather than by the local enthusiasts.[6] Greene also wrote a letter to Mrs. Noguchi telling her of the village and the family that considered her one of them.[7] A note in the Rockefeller family archives states that letters of October 1928 from Flexner and enclosures concerning Noguchi were destroyed at Rockefeller's direction in March 1930, a few months after Greene's return. No explanation has been found for this decision. As far as is known, the Rockefeller family sent no financial assistance to *Chikubakai*—certainly they never established a Noguchi prize.

Noguchi's more valuable medals were withheld from the shipment of his possessions that the institute sent to Japan shortly after his death because the facilities for taking proper care of them did not exist yet in Inawashiro. The medals were finally sent some twenty-five years later, through the American Embassy in Tokyo, in time for the celebration of Noguchi's birthday in 1955.[8]

Undeterred by the absence of Rockefeller sponsorship, the stubborn Kobayashi continued to carry out his original plans. He found support from various sources in Japan, and he received a gift of a few thousand dollars from Mrs. Noguchi when she sold the house at Shandaken.[9] Her understanding and respect for her husband's wishes, exemplified by this gift, are symbolically recognized by the only memorial to her existence. On the specially designed Japanese style headstone on the grave in Woodlawn Cemetery, the only name that appears is Hideyo Noguchi; there is no indication that his wife, too, is buried there. But in the tiny cemetery at Inawashiro, among the many Nihei and Noguchi family markers, there is a fine stone on which is carved in Japanese calligraphy, Noguchi Mary.

Monuments for Noguchi have proliferated in various places throughout the world, at least one of which was sent to Japan by Westerners. The bronze plaque and the bas relief are still in place in Ecuador; a statue was carved and erected in Yucatán; and a Japanese garden was created around a memorial stone at the time when the Gold Coast declared its independence as the State of Ghana.[10] Emanuel Libman, Noguchi's personal physician and friend, endowed the Noguchi Memorial Lectures in the School of Public Health at Johns Hopkins University.[11] As a contribution to the celebration of the eightieth anniversary of Noguchi's birth and in honor of Professor Charles Kofoid whom Noguchi had met in Jamaica, Dr. W. H. Brown joined with the University of California at Berkeley, the University of Arizona, and some Japanese-Americans in the San Francisco area to send a statue of Noguchi to Inawashiro. Brown had been a student of Kofoid's and knew that Kofoid had wanted to do this himself before he died but had been prevented from doing so by World War II.[12] In 1976 observances of the hundredth anniversary of Noguchi's birth were held in many parts of the world.

Miss Tilden sent Kobayashi the money that Noguchi had left with her for his sister in case he did not return from Africa. She was aware of rumors about her relationship with Noguchi that had circulated around the institute.[13] Perhaps Miss Tilden wondered at the time of Noguchi's death whether Flexner, too, had heard these rumors. A note that she wrote to Flexner on the day after Noguchi's death can be read as a denial of any basis for these rumors even as it suggests that Mrs. Noguchi may either have heard them or may somehow have initiated them:

> With further reference to Mrs. Noguchi, please understand that I do not bear her any grudge because of her previous unfairness, which of course was due to purely emotional causes. If I can be of any service to her now, of course I shall be glad to help, though I would prefer to wait until she asks my help, rather than to tender it.[14]

Since it was the policy to close down the laboratory of any staff member who left the institute, Miss Tilden's primary assignment was to complete Noguchi's unfinished work, but Flexner encouraged her to continue work toward her degree as well. She finished the Oroya fever work and took on much of the responsibility for continuing the trachoma work, but first she tried desperately to culture the organism Noguchi thought he had identified with yellow fever in Africa. When this effort failed, Miss Tilden concentrated on the other work and did receive her degree in 1931. When she had not been able to carry the trachoma work to a conclusion that confirmed Noguchi's discovery of the causative organism, Flexner removed her from the project and turned the work over to Dr. Peter Olitsky. The results were equally unsatisfactory in Olitsky's hands, but he gave Noguchi's name to two organisms that produced special forms of conjunctivitis in animals.[15]

Hoshi sponsored a trip for Eckstein to gather material in Japan for a biography,[16] Chiwaki sponsored the preparation of Okumura's biography with its careful editing of Noguchi's letters, and Dr. J. M. Carbo-Noboa, professor of tropical medicine at the school of medicine at Guayaquil, Ecuador, wrote a lengthy biographical tribute in Spanish.[17]

In the following years, Noguchi's work with snake venoms, syphilis, paresis, Bartonellosis, and spirochetes continued to be cited in Western textbooks, but his name is rarely recognized by later generations of physicians. In Japan, Noguchi seldom receives even this textbook recognition, but his memory is kept alive as a folk-hero, a model of filial virtues, created for schoolchildren by Kobayashi.

But as Noguchi's colleagues in the West began to write their memoirs, the recollections brought out much that had not been told before, raising questions that had lain dormant. As these scientists looked back to appraise the achievements and shortcomings of their generation, they speculated on the role of the institute, and of Noguchi as a symbol of its public facade and private failings. What were Noguchi's real attributes and deficiencies as a scientist? What role had the institute, and its dictatorial director, Simon Flexner, played in the development of American medical research in general and in Noguchi's controversial studies in particular? The discussions inevitably led to a question that had troubled them all more than they wished to admit: had Noguchi's infection with yellow fever been self-inflicted with suicidal intent, a bacteriologist's hara-kiri?[18]

A younger generation of Japanese writers also began to question the idealized picture of Noguchi that had been commonly accepted, and to ask why their national hero was presented to the Japanese people in such an unrealistic way. Tsukuba noted that the question of Noguchi's death being a suicide had been raised in the West and he attributed even the raising of the question to anti-Japanese prejudice.[19]

It appears most unlikely that anyone will ever know for certain whether Noguchi committed suicide. The only justification for invading his privacy to examine the evidence bearing on this question is a concern for the other doubts that it raises. Were the external pressures upon him so severe that they precipitated such action on his part? And whether or not his death was a suicide, the general questions raised by his life as a scientist are legitimate and worthy of answers, even hypothetical answers, which may be found in the consideration of Noguchi's private struggle with life as he found it.

Even to speculate on such matters means to formulate still other questions: what, beyond his genetic endowment, made Noguchi the driving, ambitious, and yet irresponsible charmer that he was — so subject to moods of dancing elation and furious despair; what were the real barriers to his acquiring the self-questioning, self-correcting techniques without which a scientist must inevitably fall into error?

What was there in his personality that invited such widely diverging views of him, which seemed to exist at times within the same person: the Japanese elite showering ceremonial honors while they privately maintained their mistrust and scorn, and country farmers creating both a national hero and a local tourist industry out of his sanctified image, wiping out the man as he really was? Why was the Western scientific community so eager to believe in the validity of his sensational pronouncements that they could exercise no guidance; and when the inevitable day of reckoning came, why did some feel so vindictive and personally betrayed whereas others felt constrained to defend him?

If we return to the beginning and consider a seventeen-month-old infant suddenly attacked by excrutiating pain followed by the anguish of weeks of illness, is it not reasonable to expect that such a child would be gravely retarded in acquiring the trust in people and objects of his environment that are so necessary for the mature person? Deficient in such trust, does such a child not have to meet life's problems with primitive responses? Hemmed in by incomprehensible fears with only the emotional defenses that a child of that age has acquired, should it come as a surprise to anyone when the child grows up with those defenses exaggerated to become his way of life?

Psychologists teach what ordinary parents with the eyes to see have readily observed: that little children feel omnipotent. They command and demand and can become little tyrants with no understanding, much less concern, for any rationality that might prevent their getting everything they want immediately. This recalls Flexner's words, incorporated into the instructions to Beeuwkes in Africa: Noguchi "is to be his own master and is to have all the facilities and provisions for his work he asks for, and given without delay. . . .There is nothing that he can ask for that should not be granted at once." Of course, these are Flexner's words, not Noguchi's. But they had been together a long time, and Flexner understood what Noguchi thought he needed.

Along with the commanding and demanding, the infant child believes in the power of magic; that dangers and prohibitions and deprivations can be made to disappear when the little king wishes them away. The rules do not apply to him; he can accomplish what others cannot, he can do what he wants with impunity because he wills that it be so. People who grow from infancy without modification of their omnipotent beliefs, particularly if they are naturally talented, can make such beliefs into self-fulfilling prophecies — for a while. They do it partly by demanding more of themselves — "I don't have to sleep and eat and take care of myself as others do" — and partly by demanding more of others — "I want it, therefore it shall be given." And as they have the demanding arrogance of the little child, so do they often retain the child's ability to charm some people even as they enrage others.

Everyone has lived through this phase of infantile omnipotence and has given it up with some regret. For some people, that regret becomes a gentle nostalgia for the carefree days of innocence; for others it remains a latent fury. A balanced memory of those days of power is a hard thing to achieve. As adults, we continue to adore the hero, the person who defies the rules and the odds and gets away with it; but how we hate and fear the loser! When such a person commits his ability, his enthusiasm and energy, and his ambitions and recklessness, to serve the interests of a benefactor, he can be very valuable as long as he appears to be winning.

The question of Noguchi's suicide must be viewed in such a context. It cannot be supported by anything from his background in Japanese tradition any more than it can be disproved by the arguments of his Western colleagues that he was an optimist by nature, and that he believed he had found the true cause of yellow fever in Africa and was, therefore, in a confident mood.[20] Although he had convinced both Mahaffy and Young that he was satisfied with the work he had done in Africa and was anxious to get back to New York to show it to Flexner, one suspects that Noguchi was struggling to convince himself. He wrote Flexner, "I have never been so certain of being correct as I am with this piece of work on West African yellow fever." It recalls his declaration to Carrel about his rabies studies before his trip to Europe: "I am now *absolutely* sure of my work."

Can these positive assertions be taken at face value or were they part of Noguchi's self-deceiving bravado? When he had really been sure of his conclusions regarding the motility and structure of spirochetes, Noguchi presented them with no apology for the lack of empirical data. How much more suspect, then, are his assertions that his vaccination protected him from the "South American form of yellow fever" and that his illness in Africa (which dumbfounded everyone else) rendered him immune to West African yellow fever. No matter how spiteful Sellards' motives may have been, no matter that his most vindictive-sounding paper gave Noguchi ample cause for saying that he had been "mercilessly ridiculed and pressed for action," Noguchi must have known that Sellards had presented substantial evidence that his own work had been wrong.

In West Africa Noguchi worked day and night, in complete chaos; he ordered and inoculated animals by the score, he made uncountable numbers of cultures, he refused competent help, and he kept his records secret. "I never could work with another man," he wrote Russell. Even though Noguchi had been working with infectious materials without accident for over twenty-five years, it would still take only a moment of carelessness to nick a finger ever so slightly. To designate such a momentary lapse as suicidal is going too far in order to make a point; but a man who exposes himself so patently to such highly infectious material, without even the protection afforded by gloves, exposes himself also to questions about his motives if not his sanity. His need to spare himself the humiliation of a public acknowledgment of his error was intense and he counted on his infantile omnipotence to see him through once more.

Was Noguchi aware than an accident had occurred when he called to say good-bye to Mrs. Young and thanked Dr. Young in the way that both found so moving? Was he aware of it when he asked at Lagos that his blood be checked for malaria? When Mahaffy talked with Noguchi in the hospital, the yellow fever had subsided but no one knew yet of the severe kidney damage that had occurred. Mahaffy's impression then conveys a sense of relief—as if, having finally been infected with yellow fever, Noguchi could abandon his grand delusion of omnipotence. Like other men, Noguchi was vulnerable; he had survived and he was glad. The bravado need no longer be sustained. Noguchi told Mahaffy, "We know very little about yellow fever." This would be the first time he ever said "We" in such a context.

Noguchi had made his way from the beginning by his appeal to the yearning for omnipotence in his mother as well as in all his sponsors. Oshika had been abandoned in her infancy, and her own history is a life of acting out the fantasies of omnipotence of a child, denying the limitations imposed by the reality of her circumstances. Cajoling, manipulating, demanding of herself and everyone else, she had set wildly ambitious goals for herself and her son even before he was born. When his hand was permanently maimed, Oshika seems to have been unable to think in terms of amelioration of the lifetime of rage and suffering to which he was condemned. Limited as she was to thoughts of her objectives for her family, she viewed his affliction only as Francis Bacon saw it in deformed persons: "a perpetual spur in himself to rescue and deliver himself from scorn. . .it stirreth in them industry. . .in a great wit, deformity is an advantage to rising." Oshika gave up none of her own fantasies; rather, she embellished them by nourishing her son's infantile notions of the power of his wishes.

Bacon's further description of the behavior of eunuchs may well be applied to much of Noguchi's behavior. The more recent Freudian observations of the personality whose growth was stunted in infancy, the various compensations sought by the mutilated person who perceives his deformity as a eunuch perceives his castration—in all these may be found characteristics manifested by Noguchi in his adolescent and adult years. Across the barriers of time and culture, with the limited data available, the pursuit of Noguchi's unconscious as it might have been reflected in his borrowing, his irresponsible management of money even though he tried to change, his narcissism, his relationships with women, his pushing his scientific interpretations beyond the level that his data justified, and his elations and his despondencies—such a pursuit is unlikely to produce conclusive evidence of the pyschological bases for these characteristics. One conclusion can be stated, however, with reasonable assurance. Oshika, the most important figure in his life, could see only one aspect of such psychological considerations—deformity is an advantage to rising—and he tried to make it so.

In holding this view, Oshika was in substantial agreement with

Kobayashi, who snapped up the challenge of making the filthy son of a drunken peasant into a educated man, inculcated with the old samurai virtues. Kobayashi's farewell words when Noguchi left Japan for America included the reminder that his deformity was one of his assets. And there was an observable yearning for omnipotence in Chiwaki, the rebel who defied the odds to make a profession out of a common trade, who used his family and elite school connections on behalf of a self-educated yokel. Flexner, too, had grandiose aspirations of his own and used this exotic, intense, and ambitious stranger to further these aspirations. Through this steady progression of sponsors, Noguchi rose to become identified with Rockefeller, the most ardent seeker of omnipotence of them all, a man whose wealth symbolized power for all the little people who pledged themselves to his service.

Such people might be expected to be highly competitive, and they were. For scientific investigations, however, it is one thing to define a problem in terms of the logical expansion of available knowledge; it is quite another to set a specific goal, such as the discovery of the cause of a particular disease, when one is not even sure that the basic science for achieving such a goal is available. It is just this distinction that Noguchi failed to make in his early years, and which his mentors permitted him to ignore, unchallenged. He began to understand it in his new beginning, but that was too late to avoid his errors in yellow fever.

Noguchi's whole professional career had been as much a response to his environment as it had been a reflection of his talent and his ways of compensating for the inadequacy of his training. In Madsen's laboratory he had been constructive, calm, and willing to work for whatever his efforts might bring. Back in New York Noguchi was quickly thrown into a "mental fit." Barely a year after his return from Copenhagen he was writing to Famulener: "As you know, we cannot find a new thing every day! It will be a great thing if one can do his own way without being compelled from the top." What Noguchi accomplished during those years was done in spite of his frantic activity rather than because of it.

Noguchi's isolation at the institute was a serious barrier to his maturation as a scientist. He had learned techniques and the working concepts of immunology in Philadelphia and Copenhagen, but even he still did not know what the gaps in his education were. As he established himself in isolation, Noguchi did not dare let his colleagues become aware of how little he knew. They referred to him as "Flexner's boy"; Corner writes that Flexner supervised Noguchi more closely than he did any of the other members of the institute. But Flexner was not a teacher; he was far too preoccupied with far too many things to supervise or even make competent judgments.

The real turning point for Noguchi came when Evelyn Tilden arrived in his laboratory. She introduced order, provided the liaison to the basic science that he lacked, and gave him the respect and devotion he could absorb in such large doses, enabling him to see that he was "not sufficiently prepared for research." As the atmosphere within his laboratory became freer, he had even begun to seek collaboration and help from others. Even though these changes came too late to avoid his original error in yellow fever, Noguchi could have withdrawn that work and corrected his errors. But the premature distribution of yellow fever antiserum and vaccine, with all the attendant publicity and ensuing bitterness, created irreparable damage.

The newspaper accounts of Noguchi's death that followed Flexner's carefully prepared press notices added immeasurably to the resentment that Noguchi had created himself through his demanding, egocentric treatment of the workers at Lagos. Major newspapers reported that Noguchi had discovered the difference between South American and African yellow fever in experiments performed with his own blood.[21] They expanded the reports with quotations from friends to whom Noguchi had written from Africa, and from past news items they filled in a background of his discoveries of the causative organisms of rabies, poliomyelitis, and "a dreaded blood disease." When these articles reached Lagos, the men there were furious.[22] Flexner's obituary sketch, which appeared in *Science* almost a year later, nourished this fury. He wrote that Stokes had determined the existence of a filter-passing virus in the African disease, a miserable understatment of the actual work done and already published by Stokes, Bauer, and Hudson. Flexner's only further reference to their work was to say: "Noguchi had completed his African studies which, among other things, confirmed Stokes's discovery of a virus and failed to yield *Leptospira icteroides.*"[23] Noguchi's memory could not have been more poorly served in the minds of the men at Lagos.

At the memorial service for Noguchi in December 1928, Theobald Smith spoke of Noguchi's work: "I have read again the papers on his yellow fever studies in South America. I do not see how anyone could have drawn inferences other than he did . . . he will stand out more and more clearly as one of the greatest, if not the greatest, figure in microbiology, since Pasteur and Koch."[24] This was a little excessive even for Clark who admits that he was prejudiced in Noguchi's favor: "I think even our great Theobald Smith had joined Jove and in this case had 'nodded.' "[25]

Opposing views of Noguchi kept appearing, not frequently but steadily, through the years in which the true nature of viral organisms in general, and of the yellow fever virus in particular, was discovered. The Rockefeller Foundation once again became involved in the manufacture and distribution of yellow fever vaccine during World War II, and once again it was premature. Long after Noguchi's death and Flexner's retirement, a vaccine made with human serum was released and the tragedy of serum hepatitis followed. On the cynical side, it came to be known among the military as the Rockefeller disease; on the tragic side, some soldiers died and another scientist was personally destroyed.

Greer Williams relates the story of Dr. Wilbur Sawyer, who had taken charge of the yellow fever research laboratory and who became the victim of this tragedy.[26] Max Theiler had developed a method for the study of yellow fever in the laboratory by passing the virus through the brains of mice. It produced an encephalitis in the mice that could be passed from one to another, growing more virulent with each passage; at the same time, however, it grew less virulent for monkeys to the point where it caused no disease but did produce immunity to yellow fever.

Theiler left Harvard to join Sawyer in the yellow fever research laboratory in New York, where Noguchi's old friend Bauer was Sawyer's assistant. When the wartime need for the vaccine became urgent, Sawyer made his fatal decision to release the product in its existing state of development, using human serum in the manufacture. Theiler advised against releasing the vaccine, but Bauer supported it. Similar work had been going on in Brazil, and dangerous side effects had appeared. Theiler believed that these side effects should be carefully investigated before any vaccine was used on a wide scale, but Bauer is reported to have believed that the vaccine was safe since the human serum had come from medical students at Johns Hopkins — "not from some dirty Brazilians."[27]

Theiler was awarded the Nobel Prize in 1951 for his contribution to the development of yellow fever vaccine, and Williams writes of the effect of this award on Sawyer, who died within a few months. Williams devoted three chapters of his book to sarcastic denunciation of Noguchi, repeating the hara-kiri story but calling him a clown, concluding that his death was probably not a suicide: "The more plausible theory is that in infecting himself, he did not know what he was doing."[28] But of Sawyer he wrote:

> Sawyer accepted the responsibility and swallowed the criticism, but he could not admit the fact that he had made an error in judgement. . . . He could not live with repudiation. Wilber Sawyer — "the sleepless soul that perished in his pride" — died on November 12, 1951. . . . Missing the early lesson of yellow fever research history — great men can make great mistakes — Sawyer fought the enemy, error, in the only way he knew how. He stood his ground and denied it access. He wanted to be a hero in the eyes of his fellow men.[29]

Now why was Noguchi's error and ensuing death perceived as so much less honorable than Sawyer's? For it does seem that Williams reflected the feelings of other workers in yellow fever research in relating his story in this way. Sawyer was a well-trained man, easier to know, with none of Noguchi's overprotesting arrogance, although just as ambitious. Sawyer had made no claims of sensational discoveries that could not be repeated by others. Part of Noguchi's trouble came from the fact that he had worked on so many problems and had made so many mistakes that some of his colleagues became wary of all his claims. "If Noguchi said it, don't believe it."

Professional contemporaries of Flexner's said that he would do anything if he thought it was in the best interests of the institute. The question is why Flexner considered his unrestrained support of Noguchi to be in the best interests of the institute. In truth, Flexner defended Noguchi's *L. icteroides* long after Noguchi had abandoned it himself, even after Noguchi's death.[30] Only once, in all that Flexner wrote, did he ever acknowledge any of Noguchi's real faults. In the last words that he wrote about Noguchi in his diary, Flexner noted one shortcoming, observing rather sadly at the same time that the imperfection seemed to be all too common:

> The simple truth is that N became very ambitious. He had got on very far, and he continued to exert himself to the utmost. . . . The great inner drive that N had was the extension and expression of his own personality. To an altogether remarkable degree he was an individualist in science. His work justified his individualism but his own early experiences and struggles did not enlarge the scope of his human sympathies. We are, I suppose, what we are, our desires are not necessarily satiated by success; indeed they may be enlarged by it.[31]

All of the people who worked for Rockefeller did so because he gave them the opportunity and facilities for doing the work they wanted to do. But they were always beholden, and their achievements were always perceived as a tribute to Rockefeller's generosity, perspicacity, and wisdom. Whatever legacy came to American medical research, education, and practice from this institute, no matter how brilliant and lasting, was bound to carry some qualifying hint of the pressure of the competitive atmosphere, some memory of sacrifices of self-fulfillment in creative work to the real or imagined requirements of the patron, some clue to each worker's perception of the climate in which the work was done.

As we see how these elements have survived to flourish in the medical world, we recognize that the institute, with all its failings, can hardly be held responsible for all of the political, social, and professional problems that the profession faces today. But its influence on attitudes, both inside and outside of the profession, is inescapable. As a member of the institute, Noguchi differed from his colleagues only in his greater susceptibility to the temptations it offered and the ambitions it stimulated.

In Japan, the academicians kept up appearances but privately preserved their contempt and ignored Noguchi's work as much as possible. They could do no less in public for a man who had been awarded the Imperial prize and honorary degrees from two of the Imperial universities. But their original rejection of his bid for academic recognition in 1915 was based on values that were the foundations of their culture. The class structure, which had governed the lives of the Japanese people and the very existence of the nation, could not be changed overnight. The universities carried a great social responsibility in support of the leadership that was opening the country to the industrialization that Japan had deliberately avoided, and they could not afford too rapid a break with tradition at such a high academic level.

Noguchi was not one of the Japanese elite—either by birth, by training, or in his personal behavior. If the Japanese were to survive the changes required by their new cultural imports, they had to protect their own institutions and mold them slowly to absorb these innovations. The universities, in fact, held out so long that even today a foreign trained Japanese scholar does not receive the welcome and position that a graduate of a Japanese university receives. Whether the Japanese have been right in maintaining this policy for so long is beyond our subject; perhaps for their own stability at the time they were well advised to deny Noguchi admission to a group whose conserving influence was of crucial importance to the nation undergoing such radical changes.

And so Noguchi's memory has been kept alive mostly as Kobayashi's folk hero, another testimonial to the vitality of the fairy-tale image. To look beyond the image, now that we are somewhat more free to do so, is to see a controversial man finding his way, and then losing it, in a world of complex men and forces. It cannot demean either Noguchi or his associates, as so many have feared, if we can gain some insight into contemporary problems by facing the facts of his life and times as we pay homage to his unique talents and recognize both his errors and his genuine scientific accomplishments.

Notes

Chapter 1: Origins

1. George Sansom, *A History of Japan 1615-1867*, pp. 98-99.
2. Ryōtarō Shiba, "Aizujin no Isshin no Kizuato" "Scars of the Restoration on Aizu People"], *Rekishi o Kikō Suru [A Journey through History]*, pp. 30-49.
3. Sansom, *History of Japan*, p. 186.
4. The Aizu Lord was appointed military governor of Kyoto in support of the shogunate protection of the court, which was still situated in Kyoto. It is said that at least a thousand samurai from Aizu went to Kyoto to participate in whatever events were brewing, and Seisuke was able to join the service of one of these samurai.
5. Sansom, *History of Japan*, pp. 232-33.
6. *Noguchi Hakushi to Sono Haha [Dr. Noguchi and His Mother]*.
7. Gustav Eckstein, *Noguchi*, p. 9.
8. Members of a gang ordinarily were pledged to serve for life. It is commonly believed that if they were evicted for treachery, they were frequently branded by having the end of one little finger cut off. Under special circumstances, they could be granted a release from their pledge of service.
9. Hideo Sekiyama, Managing director, Noguchi Memorial Foundation, Tokyo, 1972: personal communication.
10. Ibid.

Chapter 2: Childhood

1. Francis Bacon, "Of Deformity," *The Essays or Counsels, Civil and Moral, of Francis Ld. Verulam Viscount St. Albans.* (Mount Vernon, N.Y.: A New Edition, The Peter Pauper Press 1963), pp. 170-71.

2. Samuel Futterman, "Personality Trends in Wives of Alcoholics," *Journal of Psychiatric Social Work* 23 (October 1953): 37-41.

3. Betty Smith, *A Tree Grows in Brooklyn.*

4. L. Takeo Doi, M.D., psychiatrist, Tokyo University School of Medicine, 1971: personal communication.

5. Jacques Grunberg 1969: personal communication. A musician in New York who was married in a joint ceremony with Noguchi and his wife, Grunberg was an old man when he told me this story, but he believed it was true. Although I do not believe it was true, I do believe that Noguchi told it.

6. Ruth Benedict, *The Chrysanthemum and the Sword*, p. 267.

7. Hisaharu Tsukuba, *Noguchi Hideyo*, p. 53.

8. For a discussion of the superposition of Confucianism and Buddhism onto Shintō, the native Japanese religion, see George Sansom, Chapter VI, "Confucianism and Buddhism," in *Japan: A Short Cultural History*. The relationship of the ethical principles drawn from these sources to *Bushidō* (the way of the warrior) is discussed by Sansom in Chapter 13, "The Breakdown of Feudalism," pp. 495-502.

9. Ibid.

10. A sentimental account of this meeting is included in memoirs dictated at the time of Noguchi's death by Sakae Kobayashi to G. Kawai and translated for the Archives of the Rockefeller Institute.

11. Magoroku Ide, *"Noguchi Hideyo: Hieiyuden"* ["Hideyo Noguchi: The Biography of an Anti-hero"].

12. The original is still in the possession of the Yago family who permitted its reproduction in *The Biography of Hideyo Noguchi*, published by the Noguchi Memorial Foundation in 1963.

13. Hideo Sekiyama, 1969: personal communication.

Chapter 3: Medical Chore Boy

1. For an account of the life of an apprentice in the household of a physician-teacher in the middle of the eighteenth century, see Eiichi Kiyooka, trans., *The Autobiography of Fukuzawa Yukichi.*

2. *Curriculum Vitae*, signed H. Noguchi, December 1900, Flexner Papers.

3. Noguchi to Kobayashi, June 28, 1901, Tsurukichi Okumura, *Noguchi Hideyo*, p. 281. Omissions are cited in Mutsuo Miyase, ed. *Noguchi Hideyo no Tegami* [*Letters of Hideyo Noguchi*], p. 12.

4. Facilities Division, Railway Supervision Bureau, Ministry of Transport, Kasumigaseki, Chiyoda-ku, Tokyo, 1970: personal communication.

Chapter 4: Up to Tokyo

1. Hideo Sekiyama, 1971: personal communication.

2. Takeshi Kawakami, *Gendai Nihon Iryō Shi* [*A History of Medical Treatment in Modern Japan*], Translation provided by courtesy of James Bartholomew of Section 2, Chapter 2, Part 3 "Zaisei (sic) Gakusha no Seisui" ["Rise and Fall of Saiseigakusha."]

3. Tsurukichi Okumura, *Noguchi Hideyo*, pp. 177-78.
4. Noguchi to Chiwaki, December 18, 1897, Okumura, *Noguchi*, p. 178.
5. Noguchi to Yago, December 31, 1897, Okumura, *Noguchi*, p. 181.
6. Gustav Eckstein, *Noguchi*, pp. 56-57.
7. Noguchi to Yago, March 20, 1898, Okumura, *Noguchi*, p. 183.
8. Eckstein, *Noguchi*, p. 57.
9. Noguchi to Flexner, January 19, 1914, Flexner Papers.
10. Dr. S. Zelig Sorkin, 1970: personal communication.

Chapter 5: At the Kitasato Institute

1. James Bartholomew, *Kitasato Shibasaburō and the Acculturation of Science in Japan*, p. 35.
2. Ibid., chaps. 2 and 3.
3. Ibid., p. 148.
4. Noguchi to Yago, June 13, 1898, Tsurukichi Okumura, *Noguchi Hideyo*, p. 189.
5. Noguchi to Ishizuka, August 6, 1898, *Noguchi Hakushi Fumetsu no Seishin* [*The Immortal Spirit of Dr. Noguchi*] Noguchi Memorial Foundation.
6. Paul F. Clark, "Hideyo Noguchi, 1876-1928," *Bulletin of the History of Medicine* 33, 1 (1959): p. 2.
7. In a personal communication, Dr. L. Takeo Doi told me that his own father had come from the island of Shikoku to Tokyo in a similar way. Dr. Doi said that his father was not a particularly ambitious man, he had a good job at home, and so was not leaving for lack of opportunity. Rather, he was discouraged in his plans by the villagers, but he could not resist the call of the outside world.
8. F. G. Notehelfer, *Kōtoku Shūsui: Portrait of a Japanese Radical*, (Cambridge: Cambridge University Press, 1971). pp. 32 and 36-37.
9. Magoroku Ide, "Noguchi Hideyo: Hieiyuden" ["Hideyo Noguchi: The Biography of an Anti-hero"], pp. 227-68.
10. Okumura, *Noguchi*, p. 200.
11. Keisa Ōyama, *Hoshi Hajime Hyōden* [*Critical Biography of Hajime Hoshi*]. This biography of Noguchi's friend Hajime Hoshi tells of Hoshi's planning to come to the United States as a result of reading such books, and of other young men who did the same.
12. Flexner Diary, entry p. 87 (volume dated March 1898).
13. Simon Flexner, "Hideyo Noguchi," *Science* 69 (1929): 653-60.
14. Okumura, *Noguchi*, p. 207.
15. Noguchi to Flexner, May 30, 1899, Flexner Papers.
16. Noguchi to Flexner, May 24, 1900, Flexner Papers.

Chapter 6: Physician in Yokohama and China

1. Tsurukichi Okumura, *Noguchi Hideyo*, p. 210.
2. Hideyo Noguchi, "Since Last Fall" from *Inawashiro Alumni Bulletin*, December

1900, reproduced in Okumura, *Noguchi,* p. 224.

3. Ibid.

4. Noguchi to Flexner, May 24, 1900, Flexner Papers.

5. Citation of Noguchi by Newchwang Sanitary Commission, June 28, 1900, Flexner Papers.

6. Noguchi, "Since Last Fall," Okumura, *Noguchi,* p. 225.

7. Okumura refers to this family as Naitō to protect their identity.

8. "The Memoirs of Seisaku Noguchi" see Chapter II Note 10.

9. Noguchi Memorial Foundation, *Hideyo Noguhi,* 1967, p. 187.

10. Noguchi to Chiwaki, December 22, 1900, Okumura, *Noguchi,* p. 258.

Chapter 7: Philadelphia

1. May 22, 1928, entry, Flexner Diary.

2. Noguchi to Flexner, May 30, 1899, Flexner Papers. Noguchi to Flexner, May 24, 1900, Flexner Papers.

3. Conversation with Peyton Rous reported by Flexner in his diary December 24, 1928.

4. Paul F. Clark, "Hideyo Noguchi, 1876-1928", *Bulletin of the History of Medicine* 33, I (1959): 3.

5. Kitasato to Flexner, November 7, 1900, Flexner Papers.

6. Alfred Mirsky, 1970: personal communication.

7. Simon Flexner, "Hideyo Noguchi," *Science* 69 (1929): 653-60.

8. December 24, 1928, entry, Flexner Diary.

9. Noguchi to Chiwaki, December 22, 1900, Tsurukichi Okumura, *Noguchi Hideyo,* p. 258.

10. *Curriculum Vitae,* signed H. Noguchi, December 1900, Flexner Papers.

11. Okumura, *Noguchi,* p. 264.

12. Hisaharu Tsukuba, *Noguchi Hideyo,* p. 107.

13. Noguchi to Chiwaki, January 3, 1901, Okumura, *Noguchi,* p. 266.

14. Hans Zinsser *Microbiology,* eds. David T. Smith, Norman F. Conant, and John R. Overman. 13th ed., p. 297.

15. Simon Flexner and Hideyo Noguchi, "Snake Venom in Relation to Haemolysis, Bacteriolysis, and Toxicity," *Journal of Experimental Medicine* 6 (1902): 277.

16. Noguchi to Kobayashi, January 2, 1902, Mutsuo Miyase, *Noguchi Hideyo no Tegami,* p. 20.

17. Noguchi to Chiwaki, February 2, 1901, Okumura, *Noguchi,* p. 269.

18. Noguchi to unidentified person, May 10, 1902, Okumura, *Noguchi,* p. 271.

19. Noguchi to Chiwaki, April 21, 1901, Okumura, *Noguchi,* p. 272.

20. Noguchi to Kobayashi, April 3, 1902, Okumura, *Noguchi,* p. 269.

21. Noguchi to Chiwaki, June 10, 1901, Okumura, *Noguchi,* p. 276.

22. Noguchi to Kobayashi, June 28, 1901, Okumura, *Noguchi,* p. 281.

23. Noguchi to Chiwaki, September 10, 1901, Okumura, *Noguchi,* p. 285.

24. Ibid., September 16, 1901, Okumura, *Noguchi,* p. 278.

25. Ibid.

26. Ibid., November 21, 1901, Okumura, *Noguchi,* p. 287.

27. Ibid., December 10, 1901, Okumura, *Noguchi,* p. 293.

28. Flexner and Noguchi, "Snake Venom," p. 277.

29. Hideyo Noguchi, "Snake Venoms," in William Osler and Thomas McCrae, *Modern Medicine*, p. 247.

30. Hideyo Noguchi, *"Snake Venoms: An Investigation of Venomous Snakes with Special Reference to the Phenomena of Their Venoms,"* Carnegie Institutuion of Washington, No. 111 (1909).

31. "Clark, "Hideyo Noguchi," *Bulletin*, p. 4.

32. December 24, 1928 entry, Flexner Diary.

Chapter 8: The Lonely Foreigner

1. Noguchi to Chiwaki, December 10, 1901, Tsurukichi Okumura, *Noguchi Hideyo*, p. 293.

2. Noguchi to Kobayashi, January 2, 1902, Mutsuo Miyase, *Noguchi Hideyo no Tegami*, p. 20.

3. Ibid.

4. Ibid.

5. Ibid.

6. Ibid.

7. Noguchi's first roominghouse was at 3328 Walnut Street. On June 10, 1901, he wrote Chiwaki that he had moved to 116 S. Thirty-third Street, the implication being that he had moved because of the disappearance of money and other things from his room. When he wrote Kodama on February 11, 1902, he had returned to his old roominghouse on Walnut Street, and he went back there again after his summer in Woods Hole, remaining until he moved in with the German family. This creates some confusion about his losses. If there was ill feeling in the first roominghouse, it seems to have been forgiven. We don't know, therefore, whether he really lost the money Flexner had given him for the coat or whether he thought it had been stolen. Furthermore, the few months during which Noguchi lived at the Thirty-third Street address may have been the time when he is said to have lived with Kodama; his return to his old address after Kodama's departure would tend to support this suggestion, but there is no explanation for his not mentioning the fact that he was living with Kodama.

8. Noguchi to Kodama, February 11, 1902, Okumura, *Noguchi*, p. 297. Noguchi's letters to Kodama were translated into Japanese for reproduction by Okumura. We have not been able to see the originals, and have presented retranslations back into English.

9. Noguchi to Kodama, April 13, 1902, Okumura, *Noguchi*, p. 313.

10. Okumura, *Noguchi*, p. 315.

11. Noguchi to old friends at Chiwaki's, March 3, 1902. Okumura, *Noguchi*, p. 299.

12. Noguchi to Chiwaki, March 13, 1902, Okumura, *Noguchi*, p. 300.

13. Ibid.

14. Paul F. Clark, "Hideyo Noguchi, 1876-1928," *Bulletin of the History of Medicine* 33, I (1959): p. 3.

15. Noguchi to Chiwaki, March 13, 1902, Okumura, *Noguchi*, p. 300.

16. Noguchi to Kobayashi, April 3, 1902, Okumura, *Noguchi*, p. 309.

17. Ibid.

18. Noguchi to Kodama, July 27, 1902, Okumura, *Noguchi*, p. 333.

19. Noguchi to Kobayashi, August 23, 1902, *Okumura*, Noguchi, p. 328.

20. George W. Corner, *A History of The Rockefeller Institute 1901-1953: Origins and Growth*, p. 112.

21. E. Schelde-Moller, *Thorvald Madsen*, p. 44.

22. James Bartholomew, *Kitasato Shibasaburō and the Acculturation of Science in Japan*, pp. 108-118.

23. Noguchi to Kobayashi, October 23, 1902, Okumura, *Noguchi*, p. 338.

24. Corner, *History*, pp. 50-53.

25. Noguchi to Kobayashi, October 23, 1902, Miyase, p. 45.

26. Noguchi to Kobayashi, February 4, 1903. Okumura, *Noguchi*, p. 351.

27. Noguchi to Kobayashi, March 22, 1903. Letter is reproduced only in part by Okumura, *Noguchi*, p. 354. Omissions are reproduced by Miyase, *Noguchi*, p. 58.

28. Ibid.

29. Noguchi to Chiwaki, August 3, 1903. Okumura, *Noguchi*, p. 360.

30. Noguchi to Kobayashi, August 30, 1903, Miyase, *Noguchi*, p. 74.

31. Ibid.

32. Helen Thomas Flexner, *My Quaker Childhood.*

33. Noguchi to Kobayashi, September 22, 1903, Okumura, *Noguchi*, p. 362.

Chapter 9: At the Little Institute in Copenhagen

1. Noguchi to Kobayashi, March 22, 1903, Tsurukichi Okumura, *Noguchi Hideyo*, p. 351.

2. E. Schelde-Møller, *Thorvald Madsen*, pp. 64-65.

3. Evelyn B. Tilden, 1970: personal communication.

4. Schelde-Møller, p. 69.

5. William Bulloch, *The History of Bacteriology*, p. 224.

6. Schelde-Moller, pp. 16-18.

7. Ibid.

8. Ibid., pp. 56-58.

9. Noguchi to Flexner, December 17, 1903. Flexner Papers.

10. Ibid., February 16, 1904, Flexner Papers.

11. Noguchi to Kobayashi, February 18, 1904, Mutsuo Miyase, *Noguchi Hideyo no Tegami*, p. 88.

12. Noguchi to Chiwaki, February 22, 1904, Okumura, *Noguchi*, p. 370.

13. Ibid., April 14, 1904, Okumura, *Noguchi*, p. 374.

14. Noguchi to Kobayashi, January 1, 1904, Miyase, *Noguchi*, p. 86.

15. Noguchi to Chiwaki, February 22, 1904, Miyase, *Noguchi*, p. 86.

16. Sten Madsen, M.D., 1973: personal communication.

17. Hideo Sekiyama, 1972: personal communication.

18. Famulener letters, New York Academy of Medicine.

19. Noguchi to Famulener, February 16, 1905, Famulener Letters.

20. Noguchi to Chiwaki, September 15, 1904, Okumura, *Noguchi*, p. 378.

Chapter 10: Anxious Beginnings in New York

1. Noguchi to Famulener, September 22, 1904, Famulener Letters.

2. Ibid., undated, Famulener Letters.

3. Ibid., October 26, 1904, Famulener Letters.

4. Noguchi to Chiwaki, November 13, 1904, Tsurukichi Okumura, *Noguchi Hideyo*, p. 384.

5. Noguchi to Kobayashi, January 1, 1905, Mutsuo Miyase, *Noguchi Hideyo no Tegami*, p. 91.

6. Noguchi to Chiwaki, January 7, 1905, Okumura, *Noguchi*, p. 386.

7. Noguchi to Famulener, February 16, 1905, Famulener Letters.

8. Noguchi to Kobayashi, April 20, 1905, Miyase, *Noguchi*, p. 94.

9. Noguchi to Chiwaki, April 23, 1905, Okumura, *Noguchi*, p. 389.

10. Noguchi to Famulener, December 16, 1905, Famulener Letters.

11. William Osler and Thomas McCrae, eds., *Modern Medicine*. (Philadelphia: Lea Brothers and Co. 1907)

12. Noguchi to Famulener, December 16, 1905, Famulener Letters.

13. Ibid.

14. Hideyo Noguchi and Simon Flexner, "On the Occurrence of *Spirochaeta Pallida*, Schaudinn, in Syphilis," *Medical News* 86 (1905): 1145.

15. Okumura, *Noguchi*, p. 392.

16. Ibid., p. 395.

17. Ibid., p. 396.

18. Ibid., p. 413.

19. Ibid., p. 418.

20. Noguchi to Chiwaki, July 19, 1907, Okumura, *Noguchi*, p. 397.

21. Noguchi to Kobayashi, November 12, 1908, Okumura, *Noguchi*, p. 399.

Chapter 11: His Family and His Marriage

1. Tsurukichi, Okumura, *Noguchi Hideyo*, p. 320.

2. Noguchi to Kobayashi, April 13, 1903, Mutsuo Miyase, *Noguchi Hideyo no Tegami*, p. 65.

3. Kobayashi to Noguchi, December 7, 1906, Miyase, *Noguchi*, p. 100.

4. Noguchi to Kobayashi, July 9, 1907, Miyase, *Noguchi*, p. 101.

5. Noguchi to Chiwaki, July 31, 1907, Tsurukichi Okumura, *Noguchi Hideyo*, p. 400.

6. Ibid., August 24, 1907, Okumura, *Noguchi*, p. 406.

7. Kobayashi to Noguchi, August 8, 1908, Miyase, *Noguchi*, p. 110.

8. Ibid., May 31, 1908, Miyase, *Noguchi*, p. 108.

9. Okumura, *Noguchi*, p. 402.

10. Kobayashi to Noguchi, August 8, 1908, Miyase, *Noguchi*, p. 110.

11. Ibid., March 4, 1911, Miyase, *Noguchi*, p. 152.

12. Oshika to Noguchi, January 23, 1912. Translation was supplied by Noguchi Memorial Foundation. Original is reproduced in *Noguchi Hideyo Den*, p. 46.

13. Noguchi to Kobayashi, December 30, 1914, Miyase, *Noguchi*, p. 248.

14. Ibid., February 4, 1903, Miyase, *Noguchi*, p. 54.

15. Ibid., December 7, 1906, Miyase, *Noguchi*, p. 100. On May 13, 1910, Kobayashi used a similarly created word for his own wife, *Kabo*, meaning "mother of the family." Miyase, *Noguchi*, p. 136.

16. Noguchi to Kobayashi, September 1, 1908, Miyase, *Noguchi*, p. 113.

17. Ibid., November 12, 1909, Miyase, *Noguchi*, p. 121.

18. Kobayashi to Noguchi, December 5, 1909, Miyase, *Noguchi*, p. 124.

19. Ibid., July 13, 1916, Miyase, *Noguchi*, p. 311.

20. Noguchi to Kobayashi, April 7, 1908, Miyase, *Nogachi*, p. 105. Kobayashi to Noguchi, May 31, 1908, Miyase, *Noguchi*, p. 108.

21. Noguchi to Kobayashi, December 17, 1908, Miyase, *Noguchi*, p. 118.

22. Kobayashi to Noguchi, January 1, 1911, Miyase, *Noguchi*, p. 147.

23. Ibid., March 4, 1911, Miyase, *Noguchi*, p. 152.

24. Noguchi to Kobayashi, February 22, 1912, Okumura, *Noguchi*, p. 474. Omissions, Miyase, *Noguchi*, p. 171.

25. Ichirō Hori "Yujin Noguchi no Taibei Katei Seikatsu" ["My Friend Noguchi's Home Life in America"], *Noguchi Memorial Foundation Bulletin*, (1959).

26. Gustav Eckstein, *Noguchi*, pp. 183-84.

27. Jacques Grunberg, 1969: personal communication.

28. Eckstein says they met in New York shortly after his return from Europe, but didn't see each other again until they happened to meet accidentally on the street.

29. Bureau of Vital Statistics, State of New Jersey, Trenton, N.J.

30. Aaron J. Rosanoff: personal communication.

31. Saul Benison stated in a personal communication that this same description of Mrs. Noguchi was given him when he was collecting material for his oral history of Dr. Tom Rivers.

32. Hori, "Yujin Noguchi," p. 3.

33. Bernard Lupinek, 1970: personal communication.

Chapter 12: The Noguchi Mystique

1. Noguchi, "Snake Venoms: An Investigation."

2. Paul F. Clark, "Hideyo Noguchi, 1876-1928," *Bulletin of the History of Medicine* 33, 1 (1959): p. 4.

3. Noguchi to Chiwaki, August 3, 1903, Tsurukichi Okumura, *Noguchi Hideyo*, p. 360.

4. Ibid., April 29, 1909, Okumura, *Noguchi*, p. 408.

5. *Syphilis: A Synopsis*, Public Health Service publication, no. 1660, p. 96.

6. Hideyo Noguchi, *Serum Diagnosis of Syphilis and the Butyric Acid Test for Syphilis*. Philadelphia. J. B. Lippincott Co., 1910.

7. Claude E. Dolman, "Noguchi (Seisaku), Hideyo," *Dictionary of Scientific Biography*, pp. 143-44.

8. Dr. Kaliski died in 1966, before I was able to speak with him. His attorney, Irving B. Loonin, provided access to some material in Kaliski's papers and had some personal recollection of Kaliski's comments about Noguchi. There is some evidence that Dr. Kaliski gave Dr. Eckstein considerable help in the form of Noguchi material, some of which may have been written. There is little to be found in Dr. Kaliski's papers. Eckstein did not list his sources, and has not made his material available.

9. March 1, 1931 entry, Flexner Diary.

10. George W. Corner, *A Hisotry of The Rockefeller Institute 1901-1953: Origins and Growth*, p. 112.

11. Evelyn B. Tilden, 1969: personal communication.

12. Noguchi to Chiwaki, November 13, 1904, Okumura, *Noguchi*, p. 384 and Noguchi to Chiwaki, July 19, 1907, Okumura, *Noguchi*, p. 397.

13. Arthur H. Bryan, Charles A. Bryan, and Charles G. Bryan, *Bacteriology: Principles and Practice*, pp. 9-10. "Robert Koch. . .set forth certain criteria for establishing the etiology of diseases. These are known as *Koch's postulates* and state (1) A specific organism must always be associated with a disease; (2) it must be isolated in pure culture; (3) when inoculated into a healthy susceptible animal, it must always produce the disease; (4) it should

be obtained again in pure culture. These were long accepted as the only proof of the etiology of a disease, but with the development of serological techniques of high specificity it is now believed that an organism may be the etiological agent even if not all of Koch's requirements have been fulfilled."

14. Noguchi to Kobayashi, February 8, 1911, Okumura, *Noguchi*, p. 438.

15. Corner, *History*, p. 112.

16. Hideyo Noguchi, "A Method for the Pure Cultivation of Pathogenic *Treponema pallidum (Spirochaeta pallida),*" *Journal of Experimental Medicine* 14 (1911): 106.

17. Hans Zinsser, *A Textbook of Bacteriology* (1934).

18. ———— *As I Remember Him: The Biography of R. S.*

19. Dr. Rebecca Lancefield, 1970: personal communication.

20. Zinsser, *As I Remember Him*, p. 180.

21. H. Zinsser and J. G. Hopkins, "A Simple Method of Inducing *Spirochaeta Pallida* to Grow in Fluid Media," *Proceedings of the New York Pathological Society* 14 (1914): 77. Also, H. Zinsser J. G. Hopkins, and R. Gilbert, "Notes on the Cultivation of *T. Pallidum,*" *Journal of Experimental Medicine* 21 (1915): 213.

22. Zinsser, *A Textbook of Bacteriology*, p. 792.

23. R. R. Willcox and T. Guthe, *Treponema Pallidum*, pp. 5 and 47-8.

24. H. Noguchi, "The Spirochetes," in *The Newer Knowledge of Bacteriology and Immunology*, E. O. Jordan and I. S. Falk, eds., p. 452.

25. John C. Whitehorn, "A Century of Psychiatric Research in America," in *One Hundred Years of American Psychiatry*, J. K. Hall, ed. p. 178.

26. Aaron J. Rosanoff, "Annual Report of the Kings Park State Hospital, 1910," Rosanoff Papers.

27. H. Noguchi and J. W. Mooore, "A Demonstration of *Treponema Pallidum* in the Brain in Cases of General Paralysis," *Journal of Experimental Medicine* 17 (1913): 232.

28. Okumura, *Noguchi*, p. 442.

29. Corner, *History*, pp. 113-14.

30. Simon Flexner, "Hideyo Noguchi," *Science* 69 (1929): 653-60.

31. Noguchi to Flexner, August 16, 1910, Flexner Papers.

32. Dr. Martin Bronfenbrenner, 1969: personal communication.

33. Informant wishes to remain anonymous.

34. J. Bronfenbrenner, "A Simplified Method for Cultivating Spirochaetes on Liquid Media," *Proceedings of the Society for Experimental Biology of New York* 11 (1914): 185-87.

35. See Chapter 20, Note 19 ff.

36. Dr. Philip Reichert, 1970: personal communication.

Chapter 13: The Critical Year

1. William James, *The Will to Believe*. (Penn.: Folcroft Library Edition, 1917).

2. Hideyo Noguchi, "The Transmission of *Treponema pallidum* from the Brains of Paretics to the Rabbit," *Journal of Experimental Medicine* 61, 2 (1913): 85.

3. Theobald Smith, Professor of Comparative Pathology and Tropical Medicine at Harvard University, had been offered the director's position at the Rockefeller Institute before Flexner, but had remained at Harvard. He served, however, for many years as an active member of the Board of Scientific Directors, and in 1914 he became director of the institute's newly created Department of Animal Pathology at Princeton, N.J. His suggestion that Noguchi use bits of animal tissue in cultures came from the techniques Smith had developed for producing anaerobic conditions in his own work. See George W. Corner, *A*

History of The Rockefeller Institute 1901-1953: Origins and Growth, pp. 49-50 and Paul F. Clark, "Hideyo Noguchi, 1876-1928," *Bulletin of the History of Medicine* 33, I (1959): p. 8.

4. Clark, "Hideyo Noguchi," p. 12.
5. March 1, 1931 entry, Flexner Diary.
6. Corner, *History*, p. 72.
7. Greene to Welch, May 20, 1913. The Rockefeller Family Archives.
8. Flexner to Greene, June 16, 1913. The Rockefeller Family Archives.
9. Noguchi to Carrel, July 25, 1913, Carrel Papers.
10. Noguchi to Flexner, July 30, 1913, Flexner Papers.
11. Ibid.
12. Ibid., August 4, 1913, Flexner Papers.
13. Ibid., August 10, 1913, Flexner Papers.
14. Noguchi to Carrel, August 11, 1913, Carrel Papers.
15. Flexner to Noguchi, August 13, 1913, Flexner Papers.
16. Noguchi to Flexner, August 14, 1913, Flexner Papers.
17. Ibid., September 2, 1913, Flexner Papers.
18. March 1, 1931 entry, Flexner Diary.
19. Corner, *History*, p. 158.
20. Noguchi: European Trip, Flexner Papers.
21. Noguchi to Flexner, September 14, 1913, Flexner Papers.
22. Flexner to Noguchi, September 24, 1913, Flexner Papers.
23. Tsurukichi Okumura, *Noguchi Hideyo*, p. 449.
24. Ibid.
25. Ibid.
26. Noguchi to Chiwaki, November 2, 1913. Parts of this letter were reproduced by Okumura, *Noguchi*, p. 457, and other parts by Gustav Eckstein, *Noguchi*, pp. 213-15.
27. MacAlister to Rockefeller, October 31, 1913, The Rockefeller Family Archives and Ehrlich to Flexner, October 2, 1913, The Rockefeller Family Archives.
28. Greene to Rockefeller, November 12, 1913, The Rockefeller Family Archives.
29. Noguchi to Kobayashi, January 18, 1914, Okumura, *Noguchi*, p. 459.
30. Ibid.
31. Dr. Michael Heidelberger, 1970: personal communication.
32. Clark, "Hideyo Noguchi," p. 10.
33. Now in the possession of this writer to whom it was presented by Mrs. Peyton Rous after her husband's death.
34. Noguchi medical records, The Rockefeller University Archives.
35. Evelyn B. Tilden, 1969: personal communication.
36. Okumura, *Noguchi*, p. 489.
37. Noguochi to Flexner, January 19, 1914. Flexner Papers.
38. Noguchi medical records, The Rockefeller University Archives.
39. Okumura, *Noguchi*, p. 420.
40. *Syphilis: A Synopsis*, Public Health Service publication, no. 1660, p. 24.

Chapter 14: Return to the Orient

1. Gates to Rockefeller, Sr., April 27, 1905, Peter Collier and David Horowitz, *The Rockefellers: An American Dynasty*, p. 103.
2. Noguchi to Chiwaki, December 10, 1914, Tsurukichi Okumura, *Noguchi Hideyo*, p. 461.

Notes   291

3. Shinichi Hoshi, *Sofu Koganei Ryōsei no Ki* [*Notes on My Grandfather, Ryosei Koganei*], pp. 212-19.

4. Noguchi to Kobayashi, August 8, 1914, Okumura, *Noguchi*, p. 466.

5. Noguchi to Chiwaki, December 10, 1914, Okumura, *Noguchi*, p. 461.

6. Hoshi, *Sofu Koganei Ryōsei no Ki*.

7. Okumura, *Noguchi*, p. 480.

8. Noguchi to Chiwaki, May 22, 1915, Okumura, *Noguchi*, p. 471.

9. Noguchi to Kobayashi, July 4, 1915, Okumura, *Noguchi*, p. 478. Omissions in Mutsuo Miyase, *Noguchi Hideyo no Tegami*, p. 272.

10. Kobayashi to Noguchi, July 22, 1915, Miyase, *Noguchi*, p. 274.

11. Noguchi to Kobayashi, July 21, 1915, Miyase, *Noguchi*, p. 278.

12. Ichirō Hori, "Yujin Noguchi, no Taibei Katei Seikatsu" ["My Friend Noguchi's Home Life in America"], Noguchi Memorial Foundation *Bulletin* (1959), p. 3.

13. Noguchi to Kobayashi, September 5, 1915, Okumura, *Noguchi*, p. 485. Omissions in Miyase, *Noguchi*, p. 280.

14. Yokohama, *Asahi*, September 6, 1915.

15. Hoshi, *Sofu Koganei Ryōsei no Ki*.

16. Okumura, *Noguchi*, p. 500.

17. Magoroku Ide, "Noguchi Hideyo: Hieiyuden" ["Hideyo Noguchi: The Biography of an Anti-Hero"], pp. 227-67.

18. Ibid.

19. *Noguchi Hakushi Fumetsu no Seishin*, [*The Immortal Spirit of Dr. Noguchi*], p. 34.

20. Okumura, *Noguchi*, p. 529.

21. According to Koganei (see Note 3), Noguchi lectured in the East Auditorium at the University of Tokyo, but the lecture was sponsored by the Tokyo Medical Association and the Nisshin Medical Corporation. Following that lecture, a banquet was sponsored by some members of the medical school. Hoshi says, "Among professors, there must have been quite a few who had mixed feelings," implying that this lecture and banquet were a recognition of Noguchi; but such important details as the absence of the members of the power structure of the university made it clear that the university had not sponsored the honor.

22. Noguchi to Flexner, September 21, 1915, Flexner Papers.

23. Tetsuta Ito and Haruichiro Matsuzaki, "The Pure Cultivation of *Spirochaeta Icterohaemorrhagiae* (Inada)," *Journal of Experimental Medicine* 23 (1916): 557-62.

24. Noguchi to Flexner, November 4, 1915, Flexner Papers.

25. Collier and Horowitz, *The Rockefellers*, pp. 99-107.

26. Allan Nevins, *Study in Power: John D. Rockefeller, Industrialist and Philanthropist*, vol. 2, pp. 156-96.

27. Alice Tisdale Hobart, *Oil for the Lamps of China*.

28. Raymond B. Fosdick, *The Story of The Rockefeller Foundation*, pp. 80-92.

29. Collier and Horowitz, *The Rockefellers*, pp. 104-5.

30. Miyajima to Flexner, September 8, 1915, Flexner Papers.

31. Flexner to Miyajima, January 11, 1916, Flexner Papers.

32. March 1, 1931, entry, Flexner Diary.

33. Address delivered at the Imperial University at Kyoto, December 1915, Flexner Papers.

34. George W. Corner, *A History of The Rockefeller Institute 1901-1953: Origins and Growth*, p. 266. Eleven years later, when Dr. Tom Rivers told Flexner and Noguchi that he was preparing to publish his studies that strongly suggested that viruses are obligatory parasites on living cells, Flexner is reported to have said, "This is a free country, Rivers; you must publish what you think is right."

35. In thus establishing priority of publication over Uhlenhuth, Flexner, in effect, violated his own rules for the journal, which stated that a paper was received for publication on the

date the manuscript arrived in the journal office. He seems to have made this exception because these Japanese papers had been published previously in Japanese journals; thus, if they could have been read by Westerners, priority would have been established.

Chapter 15: Starting Over

1. Tetsuta Ito and Haruichiro Matsuzaki, "The Pure Civilization of *Spirochaeta Icterohaemorrhagiae* (Inada)," *Journal of Experimental Medicine* 23 (1916): 557.

2. Hideyo Noguchi, "Spirochaetes," *Journal of Laboratory and Clinical Medicine*, 2 no. 6 (1917): 365-400.

3. March 1, 1931, entry, Flexner Diary.

4. Gustav Eckstein, *Noguchi*, pp. 242-44.

5. Tsurukichi Okumura, *Noguchi Hideyo*, p. 585.

6. Noguchi, "Spirochaetes," pp. 365-6.

7. Noguchi to Kobayashi, April 2, 1916, Okumura, *Noguchi*, p. 540.

8. Hideyo Noguchi, "*Spirochaeta Icterohaemorrhagiae* in American Wild Rats and Its Relation to the Japanese and European Strains," *Journal of Experimental Medicine* 25 (1917): 755-63.

9. Heiser to Flexner, November 20, 1916, Flexner Papers and Victor Heiser, *An American Doctor's Odyssey*, pp. 288-90.

10. Victor Heiser, 1969: personal communication.

11. Flexner to Noguchi, August 6, 1916, Flexner Papers.

12. Noguchi to Flexner, August 9, 1916, Flexner Papers.

13. Ibid., August 24, 1916, Flexner Papers.

14. John R. Paul, *A History of Poliomyelitis*, p. 156.

15. George W. Corner, *A History of The Rockefeller Institute 1901-1953: Origins and Growth*, pp. 137-44.

16. Noguchi, "*Spirochaeta Icterohaemorrhagiae*," pp. 755-63 and Hideyo Noguchi, "A Comparative Study of Experimental Prophylactic Inoculation Against *Leptospira Icterohaemorrhagiae*," *Journal of Experimental Medicine* 28 (1918): 561.

17. Noguchi, "*Spirochaeta Icterohaemorrhagiae*," pp. 755-63.

18. Evelyn B. Tilden, 1969: personal communication.

19. Eckstein, *Noguchi*, p. 246.

20. Noguchi to Kobayashi, February 12, 1918, Okumura, *Noguchi*, p. 552.

21. Noguchi to Flexner, June 22, 1917, Flexner Papers.

22. Ibid., June 22, 1917 (2d. letter), Flexner Papers.

23. Mount Sinai prepared summaries of Noguchi's hospital records and sent them to the Rockefeller Institute. They are now a part of Noguchi's medical records in the Archives of the Rockefeller University.

24. Okumura, *Noguchi*, p. 551.

25. Noguchi to Kobayashi, February 12, 1918, Okumura, *Noguchi*, p. 551.

26. Noguchi to Flexner, April 19, 1918, Flexner Papers.

27. Evelyn B. Tilden, 1969: personal communication.

28. Ibid.

29. Noguchi and Ohira, "The Cultivation of Trichomonas of the Human Mouth (*Tetratrichomonas hominis*)" *Journal of Experimental Medicine* 25 (1917): 341.

30. Evelyn B. Tilden, 1969: personal communication.

31. Dr. Kanematsu Sugiura, 1970: personal communication. Dr. Sugiura came to the

United States as a student in 1915. Although he was younger than Noguchi, they became friends. Dr. Sugiura later specialized in cancer research at Sloane-Kettering Institute for Cancer Research, where he was still doing some work in 1970 although officially retired.

32. Noguchi to Kobayashi, June 27, 1918, Okumura, *Noguchi*, p. 556.

Chapter 16: Yellow Fever in Guayaquil

1. Noguchi to Flexner, September 18, 1918, Flexner Papers.
2. Andrew J. Warren, "Landmarks in the Conquest of Yellow Fever," *Yellow Fever*, George K. Strode, ed. 1-37.
3. Weil's disease is communicated through the pollution of water, mines, trenches, and the like, by the urine of infected rats, mice, skunks, dogs, foxes, and raccoons. The organism may enter the body through an abrasion on the skin, but is usually taken in by mouth.
4. Japanese were not granted entry into Ecuador at this time except for limited numbers who were brought in as contract laborers.
5. Noguchi to Flexner, July 21, 1918, Flexner Papers.
6. Kendall to Rose, July 19, 1918, The Rockefeller Foundation Archives.
7. Noguchi to Flexner, July 21, 1918, Flexner Papers.
8. Noguchi to Flexner, August 17, 1918, Flexner Papers.
9. Gustav Eckstein, *Noguchi*, p. 277 and Greer Williams, *The Plague Killers*, p. 220.
10. Mario Lebredo, "Considerations Suggested by Publications of Dr. Noguchi in Experimental Yellow Fever—with Appendix," *New Orleans Medical and Surgical Journal* 72 (1919), pp. 499-512.
11. Noguchi to Flexner, August 17, 1918, Flexner Papers.
12. Ibid.
13. Mario Lebredo, *Situacion epidemiologica de Guyaquil*, (1918), The Rockefeller Foundation Archives.
14. Eckstein, *Noguchi*, p. 268.
15. Charles A. Elliott, *Report on the Clinical Manifestations of Yellow Fever as Seen at Guayaquil, Ecuador*, The Rockefeller Foundation Archives.
16. Noguchi to Flexner, September 2, 1918, Flexner Papers.
17. Kendall to Rose, August 25, 1918, The Rockefeller Foundation Archives.
18. Ibid., July 30, 1918, The Rockefeller Foundation Archives.
19. Noguchi to Flexner, September 2, 1918, Flexner Papers.
20. Kendall to Rose, September 28, 1918, The Rockefeller Foundation Archives.
21. Ibid., January 28, 1919, The Rockefeller Foundation Archives.
22. Ibid., October 2, 1918, The Rockefeller Foundation Archives.
23. Flexner to Noguchi, September 12, 1918, Flexner Papers.
24. Noguchi to Flexner, September 18, 1918, Flexner Papers.
25. Noguchi to Madsen, October 26, 1918, Madsen Papers.
26. Y. Ido, H. Ito, and H. Wani, "*Spirochaeta Hebdomadis*, the Causative Agent of Seven-Day Fever (*nanukayami*)," *Journal of Experimental Medicine* 28 (1918): 435-46.
27. Noguchi to Madsen, October 26, 1918, Madsen Papers.
28. León Becerra to Flexner, March 24, 1919. The Rockefeller University Archives.
29. Hideyo Noguchi, *Report of Expedition to Ecuador as Bacteriologist of Yellow Fever Commission*, no. 7419, February 25, 1919, Flexner Papers.
30. Hideyo Noguchi, "Etiology of Yellow Fever VIII: Presence of a Leptospira in Wild Animals in Guayaquil and Its Relation to *Leptospira Icterohaemorrhagiae* and *Leptospira*

Icteroides,"Journal of Experimental Medicine 30, no. 2 (1919): 95-107.

31. Hideyo Noguchi, "Etiology of Yellow Fever II: Transmission Experiments on Yellow Fever," *Journal of Experimental Medicine* 39, no. 6 (1919): 565-84.

32. Evelyn B. Tilden, 1969: personal communication.

33. Austin Kerr, M.D., and Fred Soper, M.D., 1971: joint personal communication.

34. Noguchi, "Etiology of Yellow Fever II," 565-84.

35. Williams, *The Plague Killers*, p. 215.

Chapter 17: Mexico, Peru, and the Fatal Commitment

1. Rose to Kendall, December 6, 1918, The Rockefeller Foundation Archives.

2. Flexner Papers.

3. Flexner to Messrs. Becerra and Aguirre, May 1, 1919, Flexner Papers.

4. Noguchi, Report of Expedition to Ecuador, February 25, 1919, Flexner Papers.

5. Hideyo Noguchi, "Contribution to the Etiology of Yellow Fever," *Journal of the American Medical Association* 72, no. 3 (1919): 187-88.

6. Noguchi to Flexner, April 25, 1919, Flexner Papers.

7. Noguchi, "Etiology of Yellow Fever II: Transmission Experiments on Yellow Fever," *Journal of Experimental Medicine* 39, no. 6 (1919), pp. 565-84.

8. Minutes, Board of Scientific Directors, The Rockefeller Institute, June 1919, The Rockefeller University Archives.

9. The Society for Tropical Medicine was temporarily without a journal. The transactions of this meeting were published in various numbers of the *New Orleans Medical and Surgical Journal*. They can be found in vol. 72 (1919-20).

10. Lebredo, "Considerations Suggested by Publications of Dr. Noguchi on Experimental Yellow Fever—with Appendix," *New Orleans Medical and Surgical Journal* 72 (1919-20): 499-512. Also appeared in Spanish as "Consideraciones que sugiere lo publicado por el Dr. Noguchi sobre Etiologia de la Fiebre Amarilla," presentado al Congreso de la Sociedad Americana de Medicina Tropical, 1919, Junio 17, Atlantic City, New Jersey, *Boletín Oficial de Sanidad y Beneficencia* 26 (1921): 463.

11. Simon Flexner, Report to the Board of Scientific Directors, The Rockefeller Institute, October 19, 1913, 38-39, The Rockefeller University Archives.

12. Evelyn B. Tilden, 1970: personal communication.

13. Noguchi to Flexner, September 12, 1919, Flexner Papers.

14. Hideyo Noguchi, "Etiology of Yellow Fever X: Comparative Studies on *Leptospira Icteroides* and *Leptospira Icterohaemorrhagiae,"Journal of Experimental Medicine* 33, no. 2 (1920): 135-58.

15. Max Theiler and Andrew Watson Sellards, "The Relationship of *L. Icterohaemorrhagiae* and *L. Icteroides* as Determined by the Pfeiffer Phenomenon in Guinea Pigs," *American Journal of Tropical Medicine* 6, no. 6 (1926):383-402.

16. Noguchi to Madsen, November 28, 1919, Madsen Papers.

17. Flexner to Noguchi, November 26, 1919, Flexner Papers.

18. Kligler to Flexner, December 8, 1919, Flexner Papers.

19. Flexner to Kligler, December 12, 1919, Flexner Papers.

20. Kligler to Flexner, December 24, 1919, Flexner Papers.

21. Flexner to Kligler, January 12, 1920, Flexner Papers.

22. Flexner to Noguchi, January 12, 1920, Flexner Papers.

23. Dr. Kligler went to Palestine in the fall of 1920 and remained there until his death in 1944. He became professor of bacteriology and hygiene, Hebrew University, Jerusalem, where he was active in malaria research and in elimination of malaria from Palestine.

24. Mario Lebredo, "A Study of the Etiology of Yellow Fever in Connection with Noguchi's Leptospira Icteroides, Experimental Researches—Mérida, Yucatán," *Boletín Oficial de Sanidad y Beneficencia* 26 (1921): 509. Original version in Spanish appears in same volume.

25. Noguchi to Flexner, December 21, 1919, Flexner Papers.

26. Ibid., January 2, 1920, Flexner Papers and Noguchi to Heiser, January 6, 1920 (copy), Flexner Papers.

27. Flexner to Noguchi, January 12, 1920, Flexner Papers.

28. Kligler to Flexner, February 10, 1920, Flexner Papers.

29. Flexner to Kligler, February 21, 1920, Flexner Papers.

30. Carrel to the Royal Caroline Institute, January 14, 1920, Carrel Papers, Georgetown University. Carrel proposed Noguchi to the Nobel Prize Committee again at least four more times.

31. Lee to Rockefeller, Jr., March 27, 1920, The Rockefeller Family Archives.

32. Anonymous, "On the Trail of the Yellow Fever Germ: From the Notes of a Bystander," *American Review of Reviews and World's Work* 61, no. 4 (1920).

33. Henry Hanson, *The Pied Piper of Peru*, 53.

34. Kligler to Flexner, March 5, 1920, Flexner Papers.

35. Flexner to Kligler, March 25, 1920 (cable), Flexner Papers.

36. Kligler to Flexner, March 16, 1920, Flexner Papers.

37. Ibid., March 31, 1920, Flexner Papers.

38. Noguchi to Flexner, April 18, 1920, Flexner Papers.

39. Kligler to Flexner, May 25, 1920, Flexner Papers.

40. Kligler to Noguchi, April 29, 1920, Flexner Papers.

41. Ibid., and Noguchi to Flexner, May 5-8, 1920, Flexner Papers.

42. Noguchi to Flexner, May 5-8, 1920, Flexner Papers.

43. Ibid.

44. Kligler to Flexner, May 25, 1920, Flexner Papers.

45. Noguchi to Flexner, May 11,. 1920, Flexner Papers.

46. Noguchi to Kobayashi, August 24, 1920, Tsurukichi Okumura, *Noguchi Hideyo*, p. 594.

47. Minutes, Board of Scientific Directors, October 16, 1920, The Rockefeller University Archives.

48. Hanson, *The Pied Piper of Peru*, p. 84.

49. Flexner to Kligler, June 10, 1920, Flexner Papers.

50. Flexner to Noguchi, June 10, 1920, Flexner Papers.

51. Noguchi to Flexner, June 29, 1920, Flexner Papers.

52. Ibid.

53. Ibid.

54. Ibid., August 18, 1920, Flexner Papers.

55. Heiser to Lyster (cable), Heiser to the Rockefeller Institute (letter), August 19, 1920, The Rockefeller Foundation Archives.

56. Evelyn B. Tilden, 1969: personal communication.

57. Noguchi to Flexner, August 23, 1920, Flexner Papers.

58. Ibid., September 14, 1920, Flexner Papers.

59. Ibid., August 28, 1920, Flexner Papers.

60. Simon Flexner, *Scientific Report to the Corporation of The Rockefeller Institute*, October 15, 1920, pp. 3-10.

61. *St. Louis Post-Dispatch*, December 31, 1919.

62. Simon Flexner, "Twenty-Five Years of Bacteriology: A Fragment of Medical Research." *Science* 52, No. 1357 (1920): 615-632.

Chapter 18: The Institute in the Twenties

1. Simon Flexner, *Scientific Report to the Corporation of The Rockefeller Institute, October 15, 1920*. The Rockefeller University Archives.

2. George W. Corner, *A History of The Rockefeller Institute 1901-1953: Origins and Growth*, p. 159.

3. Ibid., p. 153.

4. Katherine Oster, 1971: personal communication.

5. Mrs. Peyton Rous, 1970: personal communication.

6. Paul de Kruif, *The Sweeping Wind*, p. 9.

7. Ibid., p. 17.

8. Anonymous, "Medicine," a chapter in *Civilization in the United States: An Inquiry by Thirty Americans*, pp. 443-45.

9. Anonymous, "Our Medicine Men, by one of them," *Century* 104 (1922); 416-26, 593-601, 701-89, 950-56.

10. Corner, *History*, p. 161.

11. De Kruif, *The Sweeping Wind*, pp. 55-57.

12. De Kruif to Archibald Mallock, April 16, 1931, Archives, New York Academy of Medicine.

13. Flexner Papers.

14. Dr. Philip McMaster, 1970: personal communication.

15. Informant wishes to remain anonymous.

16. De Kruif, *The Sweeping Wind*, p. 35.

17. Corner, *History*, p. 454.

18. Dr. John Northrop, 1972: personal communication.

19. In a personal communication, Edmund Wilson wrote: "The only thing I know about Noguchi was a statement of John Jay Chapman's—I don't remember where—that the authorities in The Rockefeller Institute didn't approve of what he was doing and stopped him by smashing his test tubes."

20. Dr. Max Delbrück, 1976: personal communication.

21. Abraham Flexner, *Medical Education in the United States and Canada*.

22. Abraham Flexner, *I Remember: The Autobiography of Abraham Flexner*, Chapters 12, and 18-22.

23. Hans Zinsser, *As I Remember Him, The Biography of R. S.*, pp. 108-11.

24. Eli Ginzberg, "Facts and Fancies About Medical Care," *Men, Money, and Medicine*.

25. Noguchi to Kobayashi, May 31, 1913, Okumura, p. 444

26. Flexner to Noguchi, January 4, 1921, Flexner Papers.

27. Northrop, on the other hand, stated that he had a budget figure he could always manage to stay within, and this was enough to satisfy Flexner. In contrast, there were not even records of the expenditures for mouse colonies in which Carrel, in his later years, tested his hypothesis of nutritional factors in the etiology of cancer. See Corner, *History*, pp. 227-29.

28. Noguchi's offer of the directorship of a new institute in Peru was the basis for his salary raise at this time. Simon Flexner, *Special Report to the Board of Directors*, October 1920, The

Rockefeller University Archives and *Salaries of Professional Staff*, the institute budget submitted to the Board of Scientific Directors for fiscal year 1927-28.

29. Dr. Kanematsu Sugiura, 1970: personal communication.
30. Informant wishes to remain anonymous.
31. Dr. Herbert L. J. Haller, 1970: personal communication.

Chapter 19: Noguchi in the Twenties

1. Dr. James S. Murphy, 1970: personal communication.
2. Gustav Eckstein, *Noguchi*, p. 284.
3. Noguchi to Kobayashi, February 18, 1919, Tsurukichi Okumura, *Noguchi Hideyo*, p. 572.
4. Okumura, *Noguchi*, p. 571.
5. Eckstein, *Noguchi*, p. 283.
6. Noguchi to Kobayashi, August 24, 1920, and August 25, 1920, Okumura, *Noguchi*, p. 599.
7. Ibid.
8. Noguchi to Kobayashi, January 31, 1921, Okumura, *Noguchi*, p. 604.
9. Hideo Sekiyama, 1969: personal communication.
10. Noguchi to Flexner, September 14, 1920, Flexner Papers.
11. Hideyo Noguchi, "Cristipira in North American Shellfish: A Note on a Spirillum Found in Oysters," *Journal of Experimental Medicine* 34 (1921): 295.
12. Hideyo Noguchi, "Leptospiras: Pathogenic and Non-pathogenic," *New York State Journal of Medicine*, 22, no. 9 (1922): 426.
13. Stokes to Flexner, November 7, 1920, Flexner Papers.
14. Mrs. Howard Cross (Ollie), 1971: personal communication.
15. Noguchi to Mrs. Cross, January 4, 1923, Noguchi Memorial Foundation.
16. Ibid., November 18, 1923. Noguchi Memorial Foundation.
17. J. Austin Kerr, 1971: personal communication. Dr. Kerr added, "Kendall was a member of the RF Yellow Fever Commission that went to Guayaquil, Ecuador, in 1918 or 1919. . . .In 1932, while on home leave in Chicago, I called on some of my professors at NUMS, including Kendall. When I told him about the disrepute into which Noguchi's leptospira had fallen, he remarked with much feeling, 'When the lion is dead, the jackals lick the corpse.' "
18. Hideyo Noguchi, "Spirochaetes (Harvey Lecture)," *Journal of Laboratory and Clinical Medicine* 2, no. 6, (1917): 365.
19. Evelyn B. Tilden, 1969: personal communication.
20. Hideyo Noguchi, *Laboratory Diagnosis of Syphilis.*
21. Okumura, *Noguchi*, pp. 581-84.
22. Noguchi to Flexner, May 22, 1922, Flexner Papers.
23. Okumura, *Noguchi*, pp. 608-12.
24. Eckstein, *Noguchi*, p. 315.
25. Noguchi to Flexner, August 12, 1922, Flexner Papers.
26. Noguchi to Wickliffe Rose, July 22, 1922, The Rockefeller Foundation Archives.
27. Okumura, *Noguchi*, p. 597.
28. Evelyn B. Tilden, 1969: personal communication.
29. Noguchi to Mrs. Flexner, March 10, 1923, Flexner Papers.
30. February 28, 1931 entry (Palermo), Flexner Diary.

31. Kobayashi to Noguchi, June 24, 1923, June 28, 1923, July 3, 1923, Miyase, pp. 542, 545, 547.

32. Noguchi to Kobayashi, September 3, 1923, Okumura, p. 614.

33. Noguchi to Mrs. Cross, November 18, 1923, Noguchi Memorial Foundation.

34. Noguchi to Flexner, June 22, 1924, Flexner Papers.

35. Noguchi to Flexner, November 28, 1924, Flexner Papers.

Chapter 20: New Work and New Critics

1. Jean Jacques Rosseau, *Émile.*

2. Report to Board of Scientific Directors, October 19, 1923, 38, The Rockefeller University Archives.

3. Hideyo Noguchi et al., *Experimental Studies of Yellow Fever in Northern Brazil,* Monographs of The Rockefeller Institute for Medical Research, No. 20 (1924).

4. J. Austin Kerr, 1971: personal communication.

5. Noguchi to Flexner, January 18, 1924, Flexner Papers.

6. Noguchi to Russell, January 31, 1924 (copy) Flexner Papers.

7. Aristedes Agramonte, M.D., "Some Observations Upon Yellow Fever Prophylaxis," *Proceedings of the International Conference on Health Problems of Tropical America, Kingston, Jamaica,* 201-27.

8. Noguchi to Flexner, August 11, 1924, Flexner Papers.

9. Agramonte, "Some Observations Upon Yellow Fever Prophylaxis," Discussion following.

10. Noguchi and Lindenberg, "The Isolation and Maintenance of Leishmania on the Medium Employed for the Cultivation of Organisms of the Leptospira Group of Spirochetes," *American Journal of Tropical Medicine* 5 (1925): 63.

11. Hideyo Noguchi, "Action of Certain Biological, Chemical, and Physical Agents upon Cultures of Leishmania: Some Observations on Plant and Insect Herpetomonads," *Proceedings of the International Conference on Health Problems in Tropical America, Kingston, Jamaica,* p. 455.

12. Ibid. Addendum and Discussion.

13. Ibid.

14. Ibid.

15. Noguchi and Tilden, "Comparative Studies of Herpetomonads and Leishmanias," *Journal of Experimental Medicine* 44 (1926) Section I: 307; Section II (by Noguchi alone): 327.

16. Hideyo Noguchi, "Abnormal Bacteria Flagella in Cultures," *Journal of the American Medical Association* 86 (1926): 1327.

17. Report to Board of Scientific Directors, October 15, 1926, 22, The Rockefeller University Archives.

18. Noguchi, "The Spirochetes," *The Newer Knowledge of Bacteriology and Immunology,* Jordan and Falk, eds. (Chicago: The University of Chicago Press 1928), pp. 452-97.

19. Hans Zinsser, *Microbiology,* eds. David T. Smith, Norman F. Conant, and John R. Overman, p. 727.

20. Noguchi, "The Spirochetes," pp. 462-63.

21. Ibid., pp. 463-64.

22. Noguchi to Flexner, March 19, 1928, Flexner Papers.

23. Noguchi, "The Spirochetes," p. 465.

24. George W. Corner, *A History of The Rockefeller Institute 1901-1953: Origins and Growth*, p. 113.

25. Noguchi, "The Spirochetes," p. 494 (explanation of Figures 19-21): p. 495 (explanation of Figures 35-42).

26. Myron G. Schultz, "A History of Bartonellosis (Carrión's Disease)," *American Journal of Tropical Medicine and Hygiene* 17, No. 4 (1968): 503-16. See also David Weinman, "Bartonellosis" in *Infectious Blood Diseases of Man and Animals*, vol. 2, ed. David Weinman and Modrac Ristic, 3-24.

27. R. P. Strong; E. E. Tyzzer; A. W. Sellards; et al., *Report of First Expedition to South America, 1913*, Harvard School of Tropical Medicine (1915).

28. Weinman, "Bartonellosis," pp. 3-24. See also Schultz, "A History of Bartonellosis," pp. 503-16.

29. Claude E. Dolman, "Noguchi, (Seisaku) Hideyo," p. 144.

30. Evelyn B. Tilden, 1969: personal communication.

31. Weinman, "Bartonellosis," pp. 3-24 and David Weinman, 1972 and 1978: personal communications.

32. M. Theiler and A. W. Sellards, "The Relationship of *L. Icterohaemorrhagiae* and *L. Icteroides.*" *The American Journal of Tropical Medicine* 6 (1926): 383-402.

33. Max Theiler, 1972: personal communication.

Chapter 21: Reality

1. Noguchi to Kobayashi, March 1, 1927, Tsurukichi Okumura, *Noguchi*, p. 628.

2. Ibid., September 8, 1924, Okumura, *Noguchi*, p. 620.

3. Noguchi to Flexner, April 19, 1925, Flexner Papers.

4. Ibid.

5. Telemaco Battistini, "The Immunological Relationships of the Leptospira Group of Spirochaetes," *Journal of Tropical Medicine and Hygiene* 28 (1925): 201-4.

6. Y. Ido, H. Ito, and H. Wani, "*Spirochaeta Hebdomadis,*" *Journal of Experimental Medicine* 28 (1918): 435-47.

7. Battistini, "The Immunological Relationships," pp. 201-4.

8. Noguchi to Flexner, August 24, 1925, Flexner Papers.

9. Ibid.

10. Hideyo Noguchi and Telemaco Battistini, "A Preliminary Report on the Cultivation of the Microbe of Oroya Fever," *Science* 63 (1926): 212.

11. Ibid., "Etiology of Oroya Fever I: Cultivation of *Bartonella Bacilliformis,*" *Journal of Experimental Medicine* 43 (1926): 851 and Ibid., "Cultivation of *Bartonella bacilliformis* from a Case of Oroya Fever," *Transactions of the Association of American Physicians* 41 (1926): 178.

12. Minutes of the Board of Scientific Directors, January 7, 1926, The Rockefeller University Archives.

13. David Weinman, 1972 and 1978: personal communications. Also David Weinman, *Infectious Anemias Due to Bartonella and Related Cell Parasites*, footnote 43, p. 270. Weinman reports that he studied the papers published by Battistini in 1925 and found that Battistini had described an organism consistent with Bartonella but that the descritpion was incomplete in several particulars.

14. Hideyo Noguchi, "Etiology of Oroya Fever II: Viability of *Bartonella bacilliformis* in Cultures and in the Preserved Blood and an Excised Nodule of *Macacus rhesus,*" *Journal of*

Experimental Medicine 44 (1926): 533.

15. TenBroeck to Theobald Smith, January 14, 1927, The Rockefeller Archive Center and Dr. John Nelson, 1970: personal communication.

16. Flexner to Russell, April 30, 1928, and May 26,1928, The Rockefeller Foundation Archives.

17. May 31, 1928, entry, Flexner Diary.

18. Dr. Charles Doan, 1970: personal communication.

19. Invitation to dinner in honor of Noguchi, Flexner Papers.

20. Remarks made at dinner in honor of Noguchi by Dr. Rufus Cole, Flexner Papers.

21. Noguchi to Flexner, January 16, 1926, Flexner Papers.

22. Noguchi to Rockefeller, January 29, 1926, The Rockefeller Family Archives.

23. Noguchi to Kobayashi, July 20, 1926, Okumura, p. 624.

24. Report to the Board of Scientific Directors, October 28, 1927, The Rockefeller University Archives.

25. Hideyo Noguchi, "Experimental Production of a Trachoma-like Condition in Monkeys by Means of a Microorganism Isolated from American Indian Trachoma," *Journal of the American Medical Association* 89 (1927): 739 and Minutes of the Board of Scientific Directors, June 3, 1927, The Rockefeller University Archives.

26.Stokes to Russell, May 15, 1927, The Rockefeller Foundation Archives.

27.Bauer to Noguchi, September 28, 1927 (copy), The Rockefeller Foundation Archives and Noguchi to Flexner, January 19, 1928, Flexner Papers.

28. W. Schüffner and Achmad Mochtar, "Versuche zur Aufteilung von Leptospiren-stämmen, mit einleitenden Bemerkungen über den Verlauf von Agglutination und Lysis," *Centralblatt für Bakteriologie, Parasitenkunde und Infectionskrankheiten* 8 (1927): 405-13 and W. Schüffner and A. Mochtar, "Gelbfieber und Weilsche Krankheit," *Archiv für Schiffs- und Tropen-Hygiene 31 (1927): 149-165.*

29. A. W. Sellards, "The Pfeiffer Reaction with Leptospira in Yellow Fever," *American Journal of Tropical Medicine* 7 (1927), 71-95.

30. A. W. Sellards and Max Theiler, "Pfeiffer Reaction and Protection Tests in Lepto-spiral Jaundice (Weil's Disease) with *Leptospira Icterohaemorrhagiae* and *Leptospira Icteroides*," *American Journal of Tropical Medicine* 7 (1927): 369-81

31. Noguchi to Flexner, August 20, 1927, Flexner Papers.

32. Ibid.

33. Noguchi to Russell, September 6, 1927, The Rockefeller Foundation Archives.

34. Russell to Beeuwkes (cable), September 14, 1927, The Rockefeller Foundation Archive

35. Paul N. Hudson, "Adrian Stokes and Yellow Fever Research: A Tribute," *Transactions of the Royal Society for Tropical Medicine and Hygiene* 60 (1966): 170-74.

36. Russell to Beeuwkes, September 23, 1927, The Rockefeller Foundation Archives.

37. Ibid., and Unlabeled document, Flexner Papers.

38. May 22, 1928, entry, Flexner Diary. The bust of Noguchi made by Mr. and Mrs. Konenkow was cast in bronze, two copies being made. One is presently in the library of the Rockefeller University. The other was given to the Japanese government and is presently in the Noguchi Memorial Hall in Tokyo.

39. Flexner to Noguchi, October 13, 1927, Flexner Papers.

40. Noguchi to Flexner, October 17, 1927, Flexner Papers.

41. Hideo Sekiyama, 1971: Personal communication.

42. Tsurukichi Okumura, *Noguchi Hideyo*, p. 646 and Gustav Eckstein, *Noguchi*, pp. 392-93.

43. Dr. Alexandre Rothen, 1970: personal communication

44. Dr. Charles Doan, 1970: personal communication.

45. Dr. Philip McMaster, 1970: personal communication.

46. Dr. Rebecca Lancefield, 1970: personal communication.

47. Dr. Philip Reichert, 1970: personal communication.
48. Flexner to Noguchi, January 3, 1928, Flexner Papers.

Chapter 22: West Africa

1. Evelyn B. Tilden, 1969: personal communication.
2. Kligler to Mrs. Kligler, July 16, 1926, Kligler Papers.
3. Ibid., August 24, 1926, Kligler Papers.
4. Cornelius B. Philip, Ph. D., 1971: personal communication; Muller to Noguchi, May 5, 1926 (copy), the Rockefeller Foundation Archives; Kligler letters to his wife from Africa, Kligler Papers.
5. Noguchi to Flexner, January 19, 1928, Flexner Papers.
6. Ibid.
7. Beeuwkes to Russell, December 26, 1927, The Rockefeller Foundation Archives.
8. Beeuwkes to Russell, March 6, 1928, The Rockefeller Foundation Archives.
9. Noguchi to Flexner, January 19, 1928, The Rockefeller University Archives.
10. Notes on Dr. Noguchi's first illness (typed copy) undated. The Rockefeller University Archives.
11. Muller to Flexner, August 25, 1928, The Rockefeller University Archives.
12. Noguchi to Flexner, January 19, 1928, The Rockefeller University Archives.
13. Russell to Noguchi, February 1, 1928, The Rockefeller University Archives.
14. Noguchi to Flexner, March 9, 1928, The Rockefeller University Archives.
15. Noguchi to Mrs. Noguchi, December 24, 1927 (copy), The Rockefeller University Archives.
16. Noguchi to Flexner, January 19, 1928, The Rockefeller University Archives.
17. Ibid.
18. Noguchi to Tilden, April 8, 1928 (copy), The Rockefeller University Archives.
19. Bauer to Noguchi, March 14, 1928 (copy), The Rockefeller University Archives.
20. Bauer to Noguchi, March 20, 1928 (copy), The Rockefeller University Archives. Miss Tilden said that Bauer wrote her regularly, using her as an outlet for his need to blow off steam. She said that she had promised him that she would never show his letters to anyone, and intended to keep her word. In a later interview, she told me that she had burned the letters. She said that they had no bearing on Noguchi, but were related to "the political situation out there."
21. Paul N. Hudson, "Adrian Stokes and Yellow Fever Research; A Tribute" *Transactions of the Royal Society for Tropical Medicine and Hygiene* 60 (1966): 170-74.
22. Noguchi to Flexner, January 19, 1929, The Rockefeller University Archives.
23. Noguchi to Tilden, April 8, 1928 (copy), The Rockefeller University Archives.
24. Johannes Bauer, "Some Characteristics of Yellow Fever Virus," *American Journal of Tropical Medicine* 11 (1931): 337-53.
25. Noguchi to Flexner, March 19, 1928, Flexner Papers.
26. Ibid.
27. Noguchi to Russell, March 25, 1928 (copy), The Rockefeller University Archives.
28. Noguchi to Flexner, April 2, 1928, Flexner Papers.
29. Max Theiler, "The Virus" Section 2 in *Yellow Fever*, George K. Strode, ed., pp. 46-47, 50-53.
30. Noguchi to Flexner, April 2, 1928, Flexner Papers.
31. A. Stokes; J. H. Bauer; and N. P. Hudson; "Transmission of Yellow Fever to *Macacus rhesus*, Preliminary Note," *Journal of the American Medical Association* 90 (1928): 253-54,

and "Experimental Transmission of Yellow Fever to Laboratory Animals," *American Journal of Tropical Medicine* 8 (1928): 103-64.

32. Greer Williams, *The Plague Killers*, 240.
33. Noguchi to Flexner, April 23, 1928 (cable), The Rockefeller University Archives.
34. Hudson to Beeuwkes, June 4, 1928 (copy), The Rockefeller University Archives.
35. Mrs. Young to Flexner, November 27, 1928, Flexner Papers.
36. May 9, 1928 entry, Beeuwkes Diary (copy), The Rockefeller University Archives.
37. Mahaffy to Flexner, June 1, 1928, The Rockefeller University Archives.
38. Russell to (at Accra), May 21, 1928, The Rockefeller University Archives.
39. May 21, 1928, entry, Beeuwkes Diary (copy), The Rockefeller University Archives.
40. Mahaffy to Russell, June 1, 1928 (copy), The Rockefeller University Archives.

Postscript

1. May 22, 1928, entry, Flexner Diary.
2. June 16, 1928, entry, Flexner Diary.
3. Letter signed "Murray" to Heiser, June 28, 1928, The Rockefeller Foundation Archives.
4. Greene to Flexner, December 24, 1929, The Rockefeller University Archives and Greene to Rockefeller, January 9, 1920, The Rockefeller Family Archives.
5. Flexner to Rockefeller, October 15, 1928, and October 17, 1928, The Rockefeller University Archives.
6. Greene to Rockefeller, January 9, 1930, The Rockefeller Family Archives.
7. Greene to Mrs. Noguchi, December 24, 1929 (copy), The Rockefeller University Archives.
8. Glenn W. Shaw (cultural attaché) to Rockefeller, October 4, 1955, The Rockefeller Family Archives.
9. Hideo Sekiyama, 1969: personal communication.
10. *Noguchi Hideyo Den* [*Biography of Hideyo Noguchi*], ed. A. Takahashi.
11. Thomas B. Turner, *Heritage of Excellence: The Johns Hopkins Medical Institutions, 1914-47*, p. 256.
12. See Note 10 above.
13. Evelyn Tilden, 1970: personal communication.
14. Tilden to Flexner, May 22, 1928, The Rockefeller University Archives.
15. Peter K. Olitsky, *Report for Oral History of Tom Rivers by Saul Benison*. Transcript of Dr. Olitsky's discussion furnished by Dr. Benison and reproduced with the consent of Mrs. Peter Olitsky.
16. Gustav Eckstein, *In Peace Japan Breeds War*, pp. 263-73.
17. J. M. Carbo-Noboa, "Hideyo Noguchi 1876-1928: Su Obra ante su tumba," *Anales de la Sociedad Medico-Quirurgica del Guayas* 8, no. 3 (1928): 101-36.
18. W. I. B. Beveridge, *The Art of Scientific Investigation*, 1950. In later editions, the observation that Noguchi had committed suicide was changed to state that it had been rumored that he had committed suicide.
19. Hisaharu Tsukuba, *Noguchi Hideyo* [*Hideyo Noguchi*], p. 232.
20. Beveridge to Rous, October 29, 1951, and December 1, 1915. Rous to Beveridge, November 26, 1951, Peyton Rous Papers.
21. *New York Times*, May 22, 1928, page 1.

22. Beeuwkes to Russell, July 21, 1928. The Rockefeller Foundation Archives.

23. Flexner, "Hideyo Noguchi," *Science* 69 (1929): 659.

24. Theobald Smith, "Memorial Address, Hideyo Noguchi, 1876-1928," *Bulletin of the New York Academy of Medicine* 5 (1929): 877-86.

25. Paul F. Clark, "Hideyo Noguchi, 1876-1928," *Bulletin of the History of Medicine* 33, I (1959): 19.

26. Greer Williams, *The Plague Killers,* pp. 320-26.

27. J. Austin Kerr, 1971: personal communication.

28. Williams, *The Plague Killers,* pp. 213-49.

29. Ibid., p. 325.

30. Max Theiler, 1972: personal communication.

31. March 1, 1931, entry, Flexner Diary.

Bibliography

Bartholomew, James. *Kitasato Shibasaburō and the Acculturation of Science in Japan*. Ann Arbor, Mich.: University Microfilms, 1972.

Beasley, William G. *The Meiji Restoration*. Stanford, Conn.: Stanford University Press, 1972.

Benedict, Ruth. *The Chrysanthemum and the Sword*. Boston, Mass.: Houghton Mifflin Co., 1946.

Benison, Saul, ed. *Tom Rivers: Reflections on a Lifetime in Medicine and Science*. Cambridge, Mass.: MIT Press, 1967.

Beveridge, W. I. B. *The Art of Scientific Investigation*. New York: W. W. Norton & Co.. 1950 (2d edition, 1954).

Binger, Carl. *The Two Faces of Medicine.*, New York: W. W. Norton & Co., 1967.

————. *Benjamin Rush, Revolutionary Doctor, 1746-1813*. New York: W. W. Norton & Co., 1966.

Bryan, Arthur H., Bryan, Charles A., and Bryan, Charles G. *Bacteriology: Principles and Practice*. New York: Barnes & Noble, 1962.

Bulloch, William. *The History of Bacteriology*. London: Oxford University Press, 1938.

Burr, Anna Robeson. *Weir Mitchell, His Life and Letters*. New York: Duffield, 1929.

Clark, Paul Franklin. "Hideyo Noguchi, 1876-1928" *Bulletin of the History of Medicine* 33, 1 (1959): 1-19.

————. *Pioneer Microbiologists of America*. Madison, Wis.: University of Wisconsin Press, 1961.

Collier, Peter, and Horowitz, David. *The Rockefellers: An American Dynasty*. New York: Holt, Rinehart & Winston, 1976.

Conroy, Hilary, and Miyakawa, Scott, eds. *East Across the Pacific*. Santa Barbara, Calif.: American Bibliographic Center-Clio Press, 1972.

Corner, George W. *A History of The Rockefeller Institute 1901-1953: Origins and Growth*. New York: The Rockefeller Institute Press, 1964.

De Kruif, Paul. *Microbe Hunters*. New York: Harcourt, Brace, and Co., 1932.

————. *The Sweeping Wind: A Memoir*. New York: Harcourt, Brace, and World, 1962.

————. "Medicine." *Civilization in the United States: An Inquiry by Thirty Americans*. Harold E. Stearns, ed., New York: Harcourt, Brace and Co., 1922.

————. "Our Medicine Men, by One of Them." *Century*. 104 (July-October, 1922): 416-426, 593-601, 781-89, 950-56.

Doi, Takeo *The Anatomy of Dependence*. Translated by John Bester. Tokyo: Kodansha International, 1973.

304

Dolman, Claude E. "Noguchi, (Seisaku) Hideyo." In *Dictionary of Scientific Biography.* Edited by Charles Coulston Gillispie, vol. 10. New York: Charles Scribner's Sons, 1974.

Eckstein, Gustav. *Noguchi.* New York: Harper and Bros., 1931.

————.*In Peace Japan Breeds War.* New York: Harper and Bros., 1943.

Fairbanks, John K., Reischauer, Edwin O., and Craig, Albert M. *A History of East Asian Civilization.* 2 vols. Boston, Mass.: Houghton Mifflin Company, 1965.

Flexner, Abraham. *I Remember: The Autobiography of Abraham Flexner.* New York: Simon and Schuster, 1940.

————.*Medical Education in the United States and Canada.* New York: Carnegie Foundation for the Advancement of Teaching, 1910.

Flexner, Helen Thomas. *A Quaker Childhood.* New Haven, Conn.: Yale University Press, 1940.

Flexner, Simon, and Flexner, James T. *William Henry Welch and the Heroic Age of American Medicine.* New York: The Viking Press, 1941.

Fosdick, Raymond B. *John D. Rockefeller, Jr.: A Portrait.* New York: Harper and Bros., 1956.

————.*The Story of The Rockefeller Foundation.* New York: Harper and Bros., 1952.

Ginzburg, Eli. *Men, Money and Medicine.* New York: Columbia University Press, 1969.

Hanson, Henry. *The Pied Piper of Peru.* Edited by Doris M. Hurnie. Jacksonville, Fl.: Convention Press, 1961.

Heiser, Victor. *An American Doctor's Odyssey.* New York: W. W. Norton & Co., 1936.

Hobart, Alice Tisdale. *Oil for the Lamps of China.* New York: Grosset and Dunlap, 1933.

Hoshi, Hajime. *Japan: A Country Founded by "Mother."* Tokyo: The Columbia University Club in Tokyo, 1937.

Hoshi, Shinichi. *Sofu Koganei Ryōsei no Ki* [*Notes on My Grandfather, Ryōsei Koganei*]. Tokyo: Kawaide Shobō Shinsha, 1974 (Japanese).

Ide, Magoroku. "Noguchi Hideyo: Hieiyuden" ["Hideyo Noguchi: The Biography of an Anti-hero"]. *Dokyumento Nihonjin* [Documents of Japanese People] vol. 1, *Kyojin Densetsu* [Legends of Giants]. Tokyo: Gakugeishorin, 1968 (Japanese).

Julianelle, L. A. *The Etiology of Trachoma.* New York: Commonwealth Fund, 1938.

Kawakami, Takeshi. "Zaisei (sic) Gakusha no Seisui" ["Rise and Fall of Saiseigakusha"]. *Gendai Nihon Iryō Shi* [*A History of Medical Treatment in Modern Japan*]. Tokyo, 1965 (Japanese), translation by James Bartholomew.

Kiyooka, Eiichi, trans. *The Autobiography of Fukuzawa Yukichi.* Tokyo: The Hokuseido Press, 1960.

LeBlanc, Thomas J. "Studies of Yellow Fever in Vera Cruz in 1920-21." *The Journal of Tropical Medicine and Hygiene* 28 no. 9, (1925): 169-78.

Marquarat, Martha. *Paul Ehrlich, 1854-1915.* London: W. Heineman Medical Books, 1949.

Miyase, Mutsuo. *Noguchi Hideyo no Tegami* [*Letters of Hideyo Noguchi*]. Tokyo: Aia Shobō, 1943 (Japanese).

Morris, Ivan. *The World of the Shining Prince.* New York: Alfred A. Knopf, 1964.

Nakane, Chie. *Japanese Society.* Berkeley, Calif.: University of California Press, 1970.

Nevins, Allan. *Study in Power: John D. Rockefeller, Industrialist and Philanthropist,* 2 vols. New York: Charles Scribner's Sons, 1953.

Noguchi Memorial Foundation. *Noguchi Hideyo Den* [*Biography of Hideyo Noguchi*]. Edited by A. Takahashi. Tokyo: 1963.

——.*Hideyo Noguchi: November 9, 1876-May 21, 1928.* Edited by A. Takahashi. Tokyo: 1967.

——.*Noguchi Hakushi Fumetsu no Seishin* [*The Immortal Spirit of Dr. Noguchi*]. Tokyo: 1955.

——.*Noguchi Hakushi to sono Haha* [*Dr. Noguchi and His Mother*]. Tokyo: 1959 (reprint, 1972).

Okumura, Tsurukichi. *Noguchi Hideyo.* Tokyo: Iwanami Shoten 1933 (Japanese).

Oliver, Wade Wright. *The Man Who Lived for Tomorrow: A Biography of William Hallock Park.* New York: E. P. Dutton, 1941.

Olson, Lawrence. *Dimensions of Japan: A Collection of Reports Written for the American Universities Field Staff.* New York: American Universities Field Staff, 1963.

Ōyama, Keisa. *Hoshi Hajime Hyōden* [*Critical Biography of Hajime Hoshi*]. Tokyo: Kyōwa Shobō, 1949 (Japanese).

Paul, John R. *A History of Poliomyelitis.* New Haven, Conn.: Yale University Press, 1971.

Penfield, Wilder. *The Difficult Art of Giving: The Epic of Alan Gregg.* Boston, Mass.: Little, Brown & Co., 1967.

Prudden, T. Mitchell. "On the Trail of the Yellow Fever Germ: From the Notes of a Bystander." *American Review of Reviews and World's Work,* 61 no. 4 (1920): 386-92.

Sansom, George. *A History of Japan, 1615-1867* vol. 3. Stanford, Conn.: Stanford University Press, 1961.

————. *Japan. A Short Cultural History,* rev. ed. New York: Appleton-Century-Crofts, Educational Division, Meredith Corporation, 1962.

Schelde-Moller, E. *Thorvald Madsen.* Copenhagen: Arnold Busck A/S, 1970 (Danish).

Schwartz, Harry. *The Case for American Medicine: A Realistic Look at Our Health Care System.* New York: David McKay Co., 1972.

Shaplen, Robert. "Toward the Well-Being of Mankind: Fifty Years of The Rockefeller Foundation." *Rockefeller Foundation Anniversary Report.* New York: Doubleday & Co., 1964.

Shiba, Ryōtarō. "Aizujin no Isshin no Kizuato" [Scars of the Restoration on Aizu People"]. *Rekishi o Kikō Suru* [*A Journey through History*]. Tokyo: Bungei Shunju, 1970: (Japanese).

Smith, Betty. *A Tree Grows in Brooklyn.* New York: Harper & Row, Publishers, 1943; 1968 (Perennial Library, Harper & Row, Publishers).

Smith, Theobald, and Welch, William H. "Hideyo Noguchi, 1876-1928." *Bulletin of the New York Academy of Medicine* 5 (1929): 877-86.

Sofue, Takao. "Kenminsei wa dō tsukurareru ka?" ["How Are Personal Characteristics Formed in the Prefectures?"]. In *Kenminsei* [Personal Characteristics in the Various Prefectures]. Tokyo: Chuō Koronsha, 1971 (Japanese).

Soupault, Robert. *Alexis Carrel, 1873-1944.* Paris: Plon, 1952 (French).

Strode, George K., ed. *Yellow Fever.* New York: McGraw-Hill Book Co., 1951.

Strong, R. P., Tyzzer, C. E., Brues, C. T., Sellards, A. W., and Gastiaburú, J. C. *Harvard School of Tropical Medicine Report of the First Expedition to South America, 1913.* Cambridge, Mass.: Harvard University Press, 1915.

Syphilis: A Synopsis. Public Health Service Publication No. 1660. Washington, D.C.: U. S. Government Printing Office, 1967.

Tsukuba, Hisaharu. *Noguchi Hideyo.* Tokyo: Kodansha, 1969 (Japanese).

Turner, Thomas B. *Heritage of Excellence. The Johns Hopkins Medical Institutions, 1914-1947.* Baltimore: Johns Hopkins University Press, 1974.

Weinman, David. *Infectious Anemias Due to Bartonella and Related Cell Parasites.* A Monograph. Philadelphia, Penn.: The American Philosophical Society, 1944.

Weinman, David, and Ristic, Modrag, eds., *Infectious Blood Diseases of Man and Animals,* 2 vols. New York: Academic Press, 1968.

Whitehorn, John C. "A Century of Psychiatric Research in America." In *One Hundred Years of American Psychiatry.* John K. Hall, ed. New York: Columbia University Press, 1944.

Willcox, R. R., and Guthe, T. *Treponema Pallidum: A Bibiliographical Review of the Morphology, Culture, and Survival of T. Pallidum and Associated Organisms.* Geneva: World Health Organization, 1966.

Williams, Greer. *The Plague Killers.* New York: Charles Scribner's Sons, 1969.

————.*Virus Hunters.* New York: Alfred A. Knopf, 1959.

Zinsser, Hans. *A Textbook of Bacteriology.* New York: D. Appleton-Century Co., 1934.

————.*As I Remember Him: The Biography of R. S.* Boston, Mass.: Little, Brown & Co., 1940.

————.*Microbiology.* Edited by David T. Smith, Norman F. Conant, and John R. Overman. 13'th ed. New York: Appleton-Century-Crofts, 1964.

Index

308